D1744132

COAL—CARRIAGE BY SEA

COAL
CARRIAGE BY SEA

By

Philip Rogers, John Strange, Brian Studd

|L|L|P|

LONDON   NEW YORK   HAMBURG   HONG KONG
LLOYD'S OF LONDON PRESS LTD
1991

Lloyd's of London Press Ltd.
Legal Publishing and Conferences Division
One Singer Street, London EC2A 4LQ

USA AND CANADA
Lloyd's of London Press Inc.
Suite 308, 611 Broadway
New York, NY 10012 USA

GERMANY
Lloyd's of London Press GmbH
59 Ehrenbergstrasse
2000 Hamburg 50, Germany

SOUTH EAST ASIA
Lloyd's of London Press (Far East) Ltd.
Room 1101, Hollywood Centre
233 Hollywood Road
Hong Kong

© Philip Rogers, John Strange, Brian Studd, 1991
First published in Great Britain, 1991

British Library Cataloguing in Publication Data
Rogers, Philip
    Coal : carriage by sea.
    I. Title II. Strange, John III. Studd, Brian
    338.2724
ISBN–1–85044–391–2

All rights reserved. No part of this publication may be reproduced,
stored in a retrieval system, or transmitted, in any form or by any means,
electronic, mechanical, photocopying, recording or otherwise, without the
prior written permission of Lloyd's of London Press Ltd.

Text set in 10 on 12pt 'Monotype' Century Schoolbook by
Megaron, Cardiff, Wales
Printed in Great Britain by
Bookcraft Ltd., Midsomer Norton, Avon

Preface

We have tried to outline the importance of coal as a fuel source and its relevance to future energy supplies. Given the decline in domestic production in many important consuming countries there is no doubt that seaborne coal trade will continue to grow over the coming decade. We hope that we have produced a practical book that gives an outline of the trade, the method that it is transported, the charter parties used, the classification and (some) of the uses that it has.

Inevitably the question of environmental considerations is raised and should be addressed. All three authors of this slim volume care deeply about the future of our planet; we seek neither to defend nor promote coal and we hope that we have reported the facts about *Coal—Carriage by Sea* in an unbiased and neutral way.

It will be clear to anyone who has read this book that coal will be with us for many years to come. In many ways it is not an ideal fuel—it is not renewable, it is dirty and expensive to handle, it gives off carbon dioxide when burnt as well as emitting particulates, sulphur dioxide (SO_2), and nitrogen oxides (NO_x). All of these problems are being addressed by many companies and official bodies (including the United Nations) around the world.

Perhaps the question that should be asked is: what are the alternatives to its use? In the long term we hope that suitable alternatives can be found, but it is totally unrealistic to imagine that all the coal-fired power stations are going to be shut down overnight. The problems of the nuclear industry are self-evident; oil is in some ways an even more precious resource with a reserve base far smaller than coal; using gas, which is an inherently far more dangerous fuel, does not solve the problem of renewability, nor the question of emissions. Renewables, such as wave and wind power, have yet to be given the green light by any industrial country in a way that contributes substantially to the country's energy needs.

The question of energy efficiency and conservation are probably the most practical measures that can be taken in the short and long term to protect the environment. There is the fear that as developing countries such as India, North and South Korea, and China push forward with their national plans they will use their natural coal reserve base to provide their energy needs, but will not have the capital to use coal efficiently to minimise its polluting effects.

Clearly the whole question of energy for the future is beyond the scope of this book, but we feel that given that coal is going to be around for so many decades to come, this book may contribute to its efficient use and transportation.

November 1991

PHILIP ROGERS
JOHN STRANGE
BRIAN STUDD

Contents

PART II CARRIAGE OF COAL

Chapter 6 The ships employed to carry coal

Chapter 7 The loading, carrying and discharging of coal

Chapter 8 Chartering

PART III ORIGINS, PROPERTIES, UTILISATION AND EVALUATION OF COAL

SECTION 1 ORIGINS, PROPERTIES AND UTILISATION OF COAL

Chapter 9 Origins

Chapter 10 The petrographic classification of coal

Chapter 11 The industrial utilisation of coal

SECTION 2 THE EVALUATION OF COALS

Chapter 12 Theoretical and practical considerations

The authors

DR PHILIP ROGERS is Director and Head of Research at Simpson Spence and Young Shipbrokers Ltd.

DR JOHN STRANGE is Senior Lecturer in the Department of Financial Services at the City of London Polytechnic.

DR BRIAN STUDD is Managing Director of Alex Stewart (Assayers) Ltd and formerly Divisional Director of Inspectorate International.

List of plates

List of figures

List of tables

PART I

Overview: coal consumption, production and trade

CHAPTER 1

Coal consumption, production and trade

COMMERCIAL TYPES OF COAL

In this book we are only considering the main commercial coals that are traded internationally by sea. This automatically excludes most sub-bituminous coals and all lignite grades (and lower ranks such as peat); it does however include anthracite although when the coal includes the latter this will be stated.

The chapters that follow refer to the individual country, or regional developments, and explain the demand for coal in the light of changes in energy demand and supply. This chapter aims to explain the overall background in terms of the major influences. In general terms these influences are perceived as steel production, electricity production, industrial and domestic demand for coal, and hard coal production.

Until more recent times (roughly the end of the 1970s) seaborne coal trade was determined by the needs of the developed countries—particularly in Europe and Japan. Since then significant additional demand has materialised in countries located in the Far East such as South Korea, Taiwan and Hong Kong. In South Korea and Taiwan (both coal producers), demand has mirrored the industrial growth in those countries over the last 30 years, particularly in relation to steel and electricity production. In these two countries a decline in domestic coal production has been in evidence over this same period. In Hong Kong (which has no natural thermal energy resources) the growth in coal demand has exactly matched the construction of new coal-fired electricity plants.

1. Energy production and general consumption

Taking 1960 as an arbitrary starting point, supply of seaborne coal has come from a variety of sources. Coal production in the developed countries has been insufficient to cover needs—irrespective of any transportation constraints. If we take the OECD grouping as representing the "developed countries", Table 1 shows that throughout the period (and even with the massive increase in supplies from the USA) these countries have been dependent upon coal from other non-OECD sources.

Traditional non-OECD coal suppliers include Poland, the USSR and South Africa. Over the past decade the People's Republic of China, Colombia, Venezuela and Indonesia have emerged as new international suppliers although, for the latter two, tonnages as yet remain relatively modest.

Table 1 OECD coal consumption, production and net imports (million tonnes oil equivalent—MTOE)

Year	Consumption	Production	Net Imports
1960	637.86	616.54	11.84
1965	668.18	660.72	18.78
1970	666.51	649.11	22.66
1975	624.78	640.71	26.83
1980	739.30	735.76	23.27
1985	816.57	773.92	19.92
1989	852.08	834.05	7.86

Source: OECD "Energy Balances of OECD Countries 1970/85" and "1980/89" International Energy Agency, Paris 1987 and 1991.

With the substantial proportion of US production (around 90%) geared to satisfying domestic demand, the pattern of growth over the last 30 years has essentially been towards Europe and Japan. This distribution also happens (conveniently) to cover the three main OECD groupings: North America (USA and Canada); Pacific (Japan, Australia and New Zealand); and Europe (the 12 countries of the European Community plus Finland, Norway, Sweden, Switzerland, Austria, Turkey and Yugoslavia).

These three areas reflect three different patterns of developments over the past 30 years:

In North America, hard coal consumption has grown rapidly, but production has risen even faster as comparison with 1960 shows. Since that time, consumption has increased by 233 MTOE, but production has increased by 302 MTOE. This surplus is available for export or is added to stocks.

Table 2 North America: coal consumption, production and net imports (MTOE)

Year	Consumption	Production	Net Imports
1960	235.43	249.38	− 14.21
1965	279.45	301.61	− 19.56
1970	308.52	351.51	− 32.87
1975	322.53	377.87	− 33.37
1980	397.45	468.42	− 56.67
1985	450.50	499.65	− 65.39
1989	468.15	550.96	− 73.78
Absolute change 1989 compared to 1960:			
	+ 232.72	+ 301.58	− 59.57

In the Pacific region there is a slightly more complicated picture in that while Japanese coal production has declined rapidly, Australian coal production has increased by more than the Japanese decline. However the demand for coal by Japan—and indeed domestic demand for coal in Australia—have combined to grow more rapidly than Australian production alone could satisfy. The result has been a net increase in production in the region, and a fast growth in consumption forcing the countries to rely on net imports.

Table 3 Pacific area: coal consumption, production and net imports (MTOE)

Year	Consumption	Production	Net Imports
1960	64.98	59.26	+ 5.33
1965	68.81	63.82	+ 8.56
1970	84.23	63.71	+ 24.47
1975	82.23	59.29	+ 23.77
1980	87.90	64.63	+ 19.00
1985	104.17	92.09	+ 8.49
1989	109.28	107.75	+ 2.14
Absolute Change 1989 compared to 1960:			
	+ 44.3	+ 48.49	− 3.19

In Europe, the period was characterised by slowly declining coal consumption, but a more rapid decline in indigenous coal production. The net effect was a shortfall of domestic supplies and a tripling of net coal imports.

Table 4 European area: coal consumption, production and net imports (MTOE)

Year	Consumption	Production	Net Imports
1960	337.45	307.90	+ 20.72
1965	319.92	295.29	+ 29.78
1970	273.76	233.89	+ 31.06
1975	220.02	203.55	+ 36.43
1980	253.95	202.71	+ 60.94
1985	260.90	182.18	+ 76.82
1989	256.64	175.34	+ 79.51
Absolute Change 1989 compared to 1960:			
	− 80.81	− 132.56	+ 58.79

2. Steel production

One of the major influences dictating the need for coal imports has been the available supplies of coking coal for the production of coke, which is then used in the blast furnace with iron ore to produce pig iron. Available supplies of good quality coking coal are distinctly limited and a substantial proportion of coal trade has been directed towards transferring this coal to the steel producing countries that did not have adequate or economic supplies of their own. The major proportion of coal trade has been directed towards Western Europe, Japan, Latin America and Asia, and in general only since the late 1970s has there been growth warranting coal imports in other areas. It is therefore relevant to consider steel production in the three major areas.

Table 5 Crude steel production (million tonnes)

Year	Western Europe	Japan	N. America
1960	109.0	22.1	95.3
1965	129.6	41.1	128.0
1970	161.5	93.3	130.5
1975	155.1	102.3	119.0
1980	161.3	111.4	117.4
1985	159.0	105.2	94.7
1990	162.4	110.3	100.8

In North America steel production at the end of the period (1990) was only slightly greater than at the start (1960). As shown in Table 2, coal production doubled over this period. Any excess capacity was therefore readily available for export. This was demonstrated in 1981 when a shortage of steam coal world wide (due to the loss of Polish supplies and Australian coal strikes) meant that an extra 20 Mt[1] of this coking and steam coal was exported at very short notice.

In the Pacific the long period of reconstruction after the Second World War in Japan was the dominant force in determining steel production and demand for coking coal. Australian steel production while increasing consistently, particularly from 1960 onwards, was not of the same order of magnitude as Japan and with adequate internal suppliers the extra coal capacity was principally destined for the Japanese market.

In Europe crude steel production reached a peak in 1974 at 186 Mt and the decline in domestic coal supplies that paralleled this increase was the major determining factor for coal imports.

3. Electricity generation

The third major influence on demand for coal imports is electricity generation and the fuel source used. It will be seen later that coal trade for electricity generation was relatively stable up to the first oil crisis (1973). At that time power station coal trade was some 10 Mt pa—out of a total trade of 104 Mt. The steam coal imports that had been seen up to that date were mainly destined for the European area and were mostly supplied by Poland.

In the OECD in 1970 coal provided 39.1% of electricity generation, by

---

1. Mt means million tons.

Table 6 Electricity from coal: OECD area (GWhr)

Year	From Coal		Market Share	Total Net Production	
1970	1,329,023		39.1%	3,402,274	
1989	2,674,459		40.3%	6,636,265	
Absolute Change	+ 1,345,436	GWhr		+ 3,233,991	GWhr

1989 it had risen only marginally to 40.3%. However these nearly identical market share figures disguise a substantial change in the absolute data.

Of this extra net electricity production, coal utilities provided 43%, while nuclear plants provided an even greater proportion: 45%, and jumped from just 2.1% market share in 1970, to 23.5% in 1989.

In North America, which is by far the largest consumer of coal for electricity (producing almost twice as much electricity by this means as OECD Europe and OECD Pacific combined) the proportion of coal used has increased significantly. Coal's proportion rose from 43% in 1970 to over 51% by 1989. At the same time oil use has declined (from 11% to 5%), while gas generation of electricity rose substantially to a peak in 1980–1981 but has since declined. Electricity generation from nuclear power plants rose 20 fold in the same period from negligible levels to 18% of total supply by 1989.

Table 7 North America: electricity production by fuel source (1,000 GWhr)

Year	Coal	Oil	Nuclear	Other	Total
1970	791	203	24	816	1,834
1975	927	320	196	846	2,289
1980	1,303	277	304	917	2,801
1985	1,580	114	467	920	3,081
1989	1,752	187	641	874	3,454

In the Pacific area in the early 1970s oil was the predominant fuel used for electricity generation accounting for over half of all production. By the end of the 1980s this share had declined to just a quarter of the region's needs. Because of the near doubling of net generation, however, this meant that in absolute terms there was only a small actual decline in the use of oil. The alternative sources used were coal, gas and nuclear plant. The growth of the latter two was quite remarkable. Between them they accounted for just 2% of electricity generation in 1970. By 1989 they were providing over 36% of a much larger base.

In Europe a near doubling of electricity generation was seen. The extra requirement of some 1,070,000 GWhr was covered mainly by the increase in new nuclear capacity. While coal-fired utilities increased their output by almost 60% because of the increase in total net production, this was not sufficient to maintain its market share which fell from 38% to 31%. Oil consumption declined but, with this exception, all other fuel sources including hydro-electricity increased their output.

Table 8 Pacific electricity production by fuel source (1,000 GWhr)

Year	Coal	Oil	Nuclear	Other	Total
1970	98	213	5	107	423
1975	94	308	25	143	570
1980	120	275	82	218	695
1985	189	200	159	272	820
1989	230	256	168	313	967

Table 9 European electricity production by fuel source (1,000 GWhr)

Year	Coal	Oil	Nuclear	Other	Total
1970	438	270	44	394	1,146
1975	437	333	111	537	1,418
1980	622	347	214	548	1,731
1985	644	188	585	567	1,984
1989	692	199	734	591	2,216

COAL TRADE

Other chapters will deal in detail with specific developments in the main exporting countries (Australia, USA, Canada, South Africa, Poland and the USSR), and the principal importing regions (Japan and Western Europe).

It will been seen that seaborne coal trade has evolved from a relatively short haul, "localised" trade centred on Europe, to a more distant truly "international" trade where the criterion of supply has been the available quantities of suitable quality coals at an acceptable price.

During the 1960s seaborne trade in coal was determined largely by developments in the steel industry. While some thermal coal was traded at the time this was a slowly declining market serving the needs of the household, domestic and minor industrial consumers. The switch away from coal in favour of more convenient and cheaper oil was quite rapid. In any event the majority of the non-coking coal traded was between countries in the European zone (including the major supplier Poland, and also the USSR).

South Africa was exporting some coal to the European market at that time, but total average annual exports were only about 1 million tonnes, of which nearly all was exported via the port of Durban and was primarily anthracite—a trade that still exists to this day. The dominant world coal suppliers were the USA (50%), Poland (14%) and the United Kingdom (11%).

Despite the overall switch to oil during the 1960s, coal trade has continued to grow virtually uninterrupted over the period. In addition, the accelerated growth in trade since 1974 reflects the effect of the oil price rises in switching from oil (eg Denmark) and either reductions in indigenous production (eg France), the increase in coal consumption, or both (eg Japan).

Table 10 Seaborne coal trade 1960–1990: major exporters (million tonnes)

Year	USA	Canada	Poland	USSR	Australia	S.Africa	UK	Other	Total
1960	22.5	0.6	6.3	3.6	1.6	1.0	5.2	5.4	46.2
1965	31.4	0.9	7.4	5.9	7.1	0.7	3.9	1.9	59.2
1970	47.5	4.0	14.7	7.7	18.3	1.3	3.2	4.5	101.2
1973	33.2	10.9	17.9	7.3	28.1	1.9	2.7	1.8	103.8
1975	44.4	11.7	22.7	8.3	29.9	2.7	2.2	5.5	127.4
1979	41.4	13.9	26.4	5.7	40.4	22.5	2.2	8.8	161.3
1980	66.1	15.3	19.3	5.3	42.8	27.9	3.8	7.3	187.8
1985	69.2	27.3	18.6	6.7	88.6	44.5	2.6	13.4	270.9
1990	81.9	29.6	17.3	7.0	106.1	49.5	2.3	47.3	341.0

PATTERN OF TRADE

While the country sections will explain national/regional developments, from a shipping point of view it is necessary to integrate these characteristics to obtain a picture of the overall pattern of trade. In this respect consideration of the various trade flows is useful in analysing this development.

Essentially the division of flows is to two major industrial areas: Europe and the Far East. Other areas such as South America, Africa and North America make relatively little impact upon the total demand pattern.

The demand for shipping has been significantly influenced by the increased length of haul of the major importing regions. This is shown below (Table 11) where it can be seen that shipping demand has increased faster than absolute trade volumes because of the increase in the distance travelled.

In practice this increase in the demand for ships (as measured in tonne–miles), reflects various changes that have developed over the years. For example, USA coal exports to Japan used to be routed almost exclusively via the Panama canal in Panamax bulk carriers. Since about 1980 an

Table 11   Coal trade index (1970 = 100)

Year	Million Tonnes	Index	Billion Tonne-Mile	Index	Miles^2	Index
1960	46.2	45.6	145	30.1	3,139	66.0
1965	59.2	58.5	216	44.9	3,650	76.8
1970	101.2	100.0	481	100.0	4,754	100.0
1973	103.8	102.6	467	97.1	4,497	94.6
1975	127.4	125.9	621	129.1	4,876	102.6
1979	161.3	159.4	795	165.3	4,930	103.7
1980	187.8	185.6	950	197.6	5,060	106.4
1985	270.9	267.7	1,490	309.7	5,500	115.7
1990	341.0	336.9	1,910	397.0	5,600	117.8

increasing proportion has been routed in Capesize (100,000 + dwt) vessels, via the Cape of Good Hope stopping-off at Richards Bay in South Africa. Equally, the decline in market share (although not necessarily in absolute terms) of both Poland and the United Kingdom as suppliers to the European market naturally implies that substitute suppliers will come from further afield.

Also, the growth in thermal coal trade has meant that South African suppliers were frequently substituted for UK and Polish coal, and even longer haul voyages, for example Australia to UK and Europe, are now commonplace.

The matrices for seaborne coal trade show the detailed breakdown of trade routes with the flows further split into coking and thermal coal [see Appendix 1].

2. Fearnley's *World Bulk Trades*, various years.

CHAPTER 2

Major coal producers

AUSTRALIA

The plentiful supply of coal of the right quality, and its relatively easy access contributed significantly to the development of the modern and vibrant coal industry that exists in Australia today. Despite its rank as the world's leading coal exporter, accounting for one-third of seaborne trade, reserves are actually much less than its export position would imply, and are estimated at just under 6%[1] of global supplies.

In total, 97.7% of reserves are located in the two main producing states; Queensland and New South Wales, and within those two states there is one main "basin". In NSW it is the "Sydney" basin, while in Queensland it is the "Bowen" basin. These two basins contain 83% of total Australian economically recoverable reserves of some 50,776 million tonnes. Part of the success of the export industry has been the ease of access to these resources and some 20% are officially classified as open-cut (open-cast) reserves.

Table 12 Australia: coal resources by type[2] (million tonnes)

Demonstrated	NSW	QLD	Other	Total
Economic	33,436	37,100	7,373	77,909

Production

Saleable coal production, which since 1980 means raw coal production less rejects removed at coal washeries plus/less unexplained colliery stock adjustments, in total was less than 17 million tonnes in 1950. Production grew slowly throughout the 1950s and was still less than 20 million tonnes in 1959. This was because of a lack of overseas opportunities and the basic orientation of the industry to serve the needs of the domestic market.

In the years since 1950, Australia's consumption of coal has more than doubled (from 17 million tonnes at that time to 50 million tonnes in 1989). Production on the other hand has increased eightfold to 155 million tonnes in 1989.[3] Although domestic stocks have built up during this time it is clear

1. World Energy Conference: *Survey of Energy Resources*, 1980.
2. *Australian Black Coal Statistics 1990*, JCB/QCB.
3. *Australian Black Coal Statistics 1989*, combined publication of the Joint Coal Board and the Queensland Coal Board.

that the growth needs of the export market have stimulated this demand. Indeed there have been many mines developed without any intention of using the coal domestically.

Table 13 Australia: saleable coal production by states (million tonnes)

Year	NSW	QLD	W.Australia	Others*	Total
1960	17.0	2.7	0.9	1.3	21.9
1965	22.7	4.3	1.0	2.1	30.1
1970	31.8	10.5	1.2	1.9	45.4
1975	34.1	22.8	2.1	1.9	60.9
1980	42.7	28.8	3.2	1.9	76.6
1985	62.3	61.0	3.8	2.3	129.4
1989	73.1	74.2	3.8	3.4	154.6

*Others includes Victoria, South Australia and Tasmania. Production ceased in Victoria in 1969

Table 14 Australia: consumption of black coal by states (million tonnes)

Year	NSW	QLD	S.Australia	W.Australia	Others*	Total
1960	13.5	2.7	1.6	1.0	1.3	20.1
1965	15.7	2.8	2.4	1.1	0.8	22.8
1970	17.0	3.6	3.0	1.2	0.3	25.1
1975	19.3	5.2	2.9	2.1	0.1	29.6
1980	23.4	6.9	2.9	3.0	0.2	36.4
1985	24.0	10.6	3.2	3.8	0.3	41.9
1989	27.5	13.2	4.1	4.3	0.4	49.6

*Others includes Victoria and Tasmania.

One of the factors encouraging efficient production has been the relentless increase in productivity, which has risen from 1.8 tonnes per employee per hour in 1980 to 3.1 tonnes in 1989. This is an average based on the output from all mines. If the underground mines are excluded, then the employee output has increased from 3.4 tonnes to 4.2 tonnes over the same period. The clear time and cost advantages associated with large open-cast operations compared to deep underground mining are readily apparent and are a further factor in making Australian coal attractive to the export market.

Consumption

It has been noted above that domestic consumption of coal has more than doubled since 1960 and there has been a persistent requirement of coking coal for use by Australia's steel industry. As the steel industry grew, in common with the general re-industrialisation of western countries after the Second World War, so too did the domestic demands for steel producers in Australia. To feed that demand there was no doubt that it was cheaper to develop indigenous metallurgical coal supplies rather than import coking coal from the nearest practical supplier, which then (as now) was the USA.

It was therefore primarily to serve the needs of the domestic market rather than the exploitation specifically for the export market, that Australia's coking coal mines were first developed. As Japan's demand for steel rapidly increased during the 1950s so too did its raw material demand for iron ore and coking coal. Besides India, Australia was the only other major iron ore supplier in the region. Supplies are available from Brazil but the distance from Brazil to Japan greatly exceeds that from Australia to Japan.

As ships were being developed for the carriage of iron ore from Australia to Japan, Japanese companies (and often the same companies) were forging links with Australian coal mining companies. There was a parallel development between Japan's needs and Australia's supply of iron ore and coking coal. The Japanese vulnerability—clearly recognised by the steel mills—was the dependence it held on these "outside" supplies. The logical step for Japanese companies was to extend their vertical integration, which they had already started doing by the use of their ships, into Australian mining companies. That ownership philosophy, which started in the 1950s, still continues to this day with many coal mines associated with Japanese firms. So it was that Japanese companies invested heavily in Australian coking coal mines and by good fortune laid the foundations for the source of supply of thermal coal that was to emerge at the end of the 1970s.

Within Australia demand for thermal coal had always been greater than that for coking coal. The major development since the war has been, like

Table 15 Australia: hard coal consumption by user (million tonnes)

Year	Power	Steel	Coke	Rail/Gas	Cement	Other	Total
1950	4.5	2.8	0.3	5.3	0.6	3.7	17.1
1955	5.5	3.4	0.3	5.1	0.8	3.4	18.6
1960	7.0	5.0	0.3	3.7	1.0	3.0	20.0
1965	10.6	6.2	0.5	2.3	0.9	2.4	22.9
1970	12.8	7.9	0.5	0.6	0.9	2.4	25.1
1975	16.4	9.0	0.4	0.1	1.0	2.6	29.6
1980	24.5	8.2	0.3	0.1	0.9	2.4	36.4
1985	30.7	6.3	0.4	0.1	1.0	3.5	41.9
1989	37.0	7.5	0.4	na	0.8	3.9	49.6

Fiscal years for 1950–1965.
Source: JCB, *Black Coal in Australia 1983/84* and *Australian Black Coal Statistics 1989*.

the situation in Europe and the United States, towards a concentration of coal use in electricity generation and the virtual elimination of its use in other sectors (like railways and gas works). This concentration in certain industries meant that while the thermal coal suppliers were facing a rather bleak future up to the mid-1970s, the surplus capacity that they had was successfully exploited to serve the needs of the new coal-fired power stations at home and, most crucially, abroad. It is clear that Australia has been extremely fortunate in terms of proximity to market, available resources, expanding mining companies, and growth in one sector as another faced decline.

Table 16   Australia: hard coal supply/demand balance (million tonnes)

Year	Production	Exports	Consumption	Stocks
1970	45.2	18.3	25.1	8.7
1975	60.7	29.9	29.6	11.7
1980	76.6	42.8	36.4	13.4
1985	129.4	87.9	41.9	26.5
1989	154.6	98.7	49.6	27.9

Source: Joint Coal Board and Queensland Coal Board

NORTH AMERICA

1. USA

First industrial growth: 1850–1920

The United States of America has long been a major producer, consumer and exporter of coal. The spread of the industrial revolution to the United States in the first half of the nineteenth century lead to increasing demands for energy supplies for power generation in factories and for transportation purposes on the railways and steamships. Until this transition phase began, wood was the prime source of energy. In 1850 for example, coal accounted for 9.3% of total fuel consumption with 219 trillion BTU, and at the same time accounted for all of the non-wood fuel consumption.[4]

The expansion of the railroads encouraged the switch away from wood in three ways: first, by providing a more concentrated form for use by the locomotives in steam raising; secondly, because the railways were hauling coal from the mines to the place of consumption and it was neither sensible

4. S.H.Schurr, B.C.Netschert *et al.*, *Energy in the American Economy 1850–1975* (Baltimore: Johns Hopkins Press for Resources for the Future, 1960).

nor economic to bring wood to the coal mines to use as a fuel, and thirdly, the iron and steel industry began to use coke rather than charcoal to reduce iron ore to pig iron.[5]

Coal utilisation grew strongly from 1850 onwards, when production was less than 8 million tonnes,[6] as the industrial base of the country expanded. The demands made on industry during the First World War pushed production to its first major peak in 1918 when output reached 525.6 million tonnes, a level that bears comparison with that of the early 1970s when output achieved a similar level (1973: 536.8 m. tonnes).

The Depression years: 1920–1939

Consequent to the artificial boost given to the coal mining industry by the First World War, demand remained high and productive capacity, which mainly consisted of many small mines, continued to expand. The effect of the post-war recession, which began in 1921 was the over-capacity and price cutting that was much in evidence.[7] Additionally, the share of US non-wood energy provided by coal, which was still in excess of 90% in 1900 began to fall more sharply from 81.6% in 1920 to 61.3% in 1930,[8] as other sources, principally oil and hydro-electricity, became available.

For the people concerned with the industry itself there was still an underlying mood of optimism although edged with the realisation that the "Golden Age" of coal was coming to an end. In his annual review of the US coal market one leading commentator[9] wrote of the year 1922:

"In past years we were accustomed to notice a doubling of the output every decade. It will be seen, therefore, how far short of that development we have fallen. It has been said that the 1922 figures compare more favourably with 1921 than might appear on the surface, on account of more coal being used at home by reason of smaller exports. But the fact remains that depressed manufacturing conditions during a good part of the year, economics and substitutions, new methods and principals, have served to put the brakes on the bituminous coal trade and suspended perhaps forever, the doubling of tonnage every ten years, which it was recognized would have come to an end sooner or later. At the same time, some moderate growth is only natural and proper, and it is thought that an upward trend will soon develop again. Even though we may never see a yearly increase of ten per cent, for no doubt economics have played their part to a large extent, yet the country will continue to grow."

Through the latter half of the 1920s and all of the 1930s, a long period of depression served to keep the US coal industry at a relatively quiet level

5. E. Tukenmez, M.K.Paull, *Outlook for U.S.Coal.* (Washington, EIA, U.S. Department of Energy, 1982).
6. J.W. Leonard, *Coal*, (Washington, National Coal Association, 1979).
7. Tukenmez & Paull *op. cit.*
8. Schurr & Netschert *op. cit.*
9. F.W. Saward, *Sawards Annual, 1923.* (New York: 1923).

although it should not be forgotten that it was still the world's leading producer throughout the whole of this time.

The post-war industrial phase: 1940–1973

The industrial demand for energy generated by the special requirements of the Second World War was another significant phase for the coal mining industry. Within the space of seven years, from 1938 to 1944, production leapt from 316 million tonnes to 562 million tonnes and almost certainly would have gone higher were it not for strikes and the manpower and equipment shortages caused by the war. In the event production was not sufficient to meet demand and other fuels, principally oil, made up the shortfall.

The year 1947 marked the start of the period when, with the exception of the three years 1920–1922, the US became and has remained ever since, a net importer of oil. The period from this time has been notable for the declining trend of coal use for transportation and a positive trend in its use by electric utilities. Coal production since 1973 has been on a slowly increasing upward trend as the US has sought to become less reliant on imported oil.

Table 17 USA: coal production, consumption, export and imports (million tonnes)

Year	Production	Consumption	Exports	Imports
1870	18.6	18.9	0.1	0.5
1880	46.1	46.4	0.2	0.5
1890	100.7	100.7	1.2	0.9
1900	192.3	186.7	5.5	1.7
1910	378.3	369.2	10.6	1.6
1920	515.9	461.8	34.9	1.1
1930	424.1	412.8	14.4	0.2
1940	418.0	391.0	14.9	0.3
1950	468.4	411.9	23.1	0.3
1960	376.9	344.7	33.1	0.2
1970	547.0	468.1	64.4	–
1980	752.7	637.5	83.2	1.1
1985	801.6	742.1	84.1	1.8
1990	933.5	811.5	96.0	2.4

Source: 1870–1970[10]; 1980–1990[11]

Developments in the US coal industry since 1973

The events relevant to the coal industry that occurred in late 1973 early

10. R.S.Manthy, *Natural Resource Commodities—A Century of Statistics, 1870–1973.* (Baltimore, Johns Hopkins UP).

11. Energy Information Administration "Monthly Energy Review" June 1991, US Dept of Energy, Washington DC USA.

1974 can be summarised as follows: On 6 October Saudi Arabian light marker crude had a posted price of $3.01 per barrel, and Syria and Egypt invaded Israel. On the 16 October, the 13 OPEC country representatives in Kuwait raised the marker price to $5.12 per barrel. On 20 October the seven OPEC countries ordered an oil embargo against the US following President Nixon's request to Congress for $2.2 billion for arms for Israel. By early November crude oil production by Arab member nations was cut to 75% of the level prevailing two months earlier. On 24 December the OPEC marker price was raised to $11.65 per barrel.[12]

In 1970, nuclear generation was starting to make a significant impact producing electricity equal to 5.7 million tonnes of oil equivalent (MTOE) out of a total primary energy consumption of 1,660 MTOE. With a well established construction programme stretching throughout the decade, prospects for further coal expansion rested largely on perceptions of future steel demand, which at that time (1970) had an expanding future ahead of it.

To summarise the position characterising the 1970–1973 period: oil was cheap and readily available, nuclear energy was starting to have an impact, natural gas was plentiful and investment in coal mining was low. Public interest in the future of coal had been suggested as being of "little concern outside the coal and railroad industries".[13]

Table 18 USA: primary energy consumption (MTOE)

Year	Oil	Gas	Coal	Nuclear	Hydro	Total
1960	396.5	323.7	240.7	0.0	12.9	973.8
1970	694.6	564.1	329.5	5.7	65.9	1,659.8
1973	818.0	572.3	335.0	21.8	75.6	1,822.7
1975	765.9	498.4	322.9	44.4	80.2	1,711.8
1979	868.0	516.4	380.7	69.3	79.8	1,914.2
1980	794.1	509.9	385.6	68.5	78.0	1,836.0
1985	721.7	445.9	437.0	103.7	84.1	1,792.3
1990	778.9	490.5	476.5	156.0	72.0	1,973.9

Source: 1960;[14] 1970–1974;[15] 1975–1990[16]

USA coal consumption by market sector

Since 1960, domestic coal consumption, ie excluding exports, has risen steadily—from 361 million tonnes in that year to a record 812 million

12. C. Goodwin, *et al*: *Energy Policy in Perspective: Today's Problems Yesterday's Solutions.* (Washington, Brookings Institution, 1981).
13. *Ibid.*, Tukenmez & Paull *op. cit.*
14. UN, "World Energy Supplies, 1960–63", Series J, No.8 (New York, 1965).
15. British Petroleum Company, *BP Statistical Review of World Energy* (London, 1983).
16. British Petroleum Company, *BP Statistical Review of World Energy* (London June, 1991).

tonnes in 1990.[17] Substantially all of that growth over the last three decades has been accounted for by increased utilisation in coal-fired utilities for the production of electricity.

Table 19   USA: percentage share of US primary energy consumption

Year	Oil	Nat. Gas	Coal	Nuclear	Hydro	Total
1960	40.7	33.2	24.7	—	1.3	100.0%
1970	41.8	34.0	19.9	0.3	4.0	100.0%
1973	44.9	31.4	18.4	1.2	4.1	100.0%
1975	44.7	29.1	18.9	2.6	4.7	100.0%
1979	45.3	27.0	19.9	3.6	4.2	100.0%
1980	43.1	27.5	21.3	3.7	4.3	100.0%
1985	40.2	24.7	24.6	5.8	4.6	100.0%
1990	39.5	24.9	24.1	7.9	3.6	100.0%

The growth in coal-fired demand in the 1960–1973 period of "cheap" oil, reflected electricity consumption (from all fuels) increasing at a faster rate than coal-fired electricity production. In 1965, for example, electricity produced from coal consumed 5,834 trillion BTU, while in 1970 this had increased to 7,483 trillion BTU, or 28% more in five years. However coal use, as a percentage of total gross energy input to the electricity generating sector, fell from 54% in 1965 to 44% by 1970.[18]

The decline in coal utilisation in the industrial sector continued rapidly after the Second World War: in 1947 coal accounted for 57% and 50%, in the industrial and household sectors, respectively. In the transportation sector 35% of energy was provided by coal, mainly for use by the railroads. By 1976, coal use in these three sectors had fallen to 21%, 1% and nil.[19] This was due to industry switching to cheaper oil, natural gas supplies and nationally supplied electricity. In the household sector switching was also seen towards oil and gas while the railroads turned to diesel fuel and reduced their consumption from 56 million tonnes in 1950 to less than 100,000 tonnes by 1973.

The United States steel industry has experienced a period of modernisation in the 1980s which, to some extent, has meant reductions in capacity mainly in the integrated sector although this has been partially offset by capacity increases in the non-integrated scrap-based sector. In addition, to this modernisation other changes have occurred in the last decade which directly affect the demand for coking coal.

In terms of steel production by process, electric furnace utilisation has

17. US DOE "Monthly Energy Review" (June 1991, Washington).

18. W.G.Dupress Jnr., J.A.West *United States Energy through the Year 2000* (Washinghton, US Govt. Printing Office, 1972).

19. US Bureau of Mines *Annual Energy Press Release.* (Washington, 1976).

Table 20 USA: coal consumption by end-use sector (million tonnes)

Year	Utilities	Coke	Industrial	Other	Total
1960	160.2	73.8	89.9	37.1	361.1
1965	222.1	86.5	96.3	23.3	428.2
1970	290.5	87.5	82.1	14.6	474.6
1973	353.1	85.4	61.9	10.1	510.4
1975	368.3	75.8	57.8	8.5	510.4
1979	478.2	70.2	61.4	7.6	617.3
1980	516.5	60.5	54.7	5.9	637.5
1985	629.4	37.2	68.4	7.1	742.1
1990	700.1	36.1	69.2	6.1	811.5

Sources: 1960–1980 US Dept of Energy, "Quarterly Coal Report". 1981–1990 US Dept of Energy, "Monthly Energy Review" June 1991.

risen from 15% of total output in 1970 to 37% in 1990.[20] Also the continuous casting ratio, which in 1970 was less than 4%, had by 1989 risen to 65%.[21] Neither of these factors directly affects the demand for coal in the production of pig iron, but both have had a major effect on overall demand for metallurgical coal.

Electric furnace operation uses steel scrap and sometimes pig iron to produce steel. As the number of these units has increased with the growth in the availability of scrap so the quantity of pig iron, and by implication coke and coking coal, has fallen. At the same time, much of the open-hearth capacity has been converted to basic oxygen furnace (bof), which has some further scope for the reduction of coke usage.

Continuous casting operations, as mentioned previously, do not directly impinge upon demand for coking coal but, by increasing the product yield from a given quantity of steel produced, the steel is produced more efficiently and the quantity of steel required is reduced.

2. Canada

Although Canada has a long tradition of exporting coal stretching back through most of this century, most of these tonnages were of small quantities only, and it was not until as recently as 1978 that the country first became a net exporter of coal. During the 1960s, coal exports averaged about 1.3–1.4 Mt annually (1 million tonnes being exported for the first time in 1964). During the rest of that decade there was little development, due largely to the location of the coal fields and the huge distances from the export ports. Towards the end of the decade however, coal producers began to market their product more aggressively, and Japanese steel companies, seeking alternative supplies to their ever increasing demands, became

20. IISI *Crude Steel production by Process* (Brussels, 1976).
21. IISI *Steel Statistical Yearbook 1990* (Brussels 1990).

more actively involved. The result of this effort was the signing, in 1968, of a long-term contract with the Japanese steel mills for the supply of 4–5 Mt annually of metallurgical coal.[22]

In fact the contract, signed on 22 March 1968, called for the supply of 5.15 Mt in fiscal 1970, and 4.4–4.5 Mt annually for 15 years after that up to 1984. The coal is known as Balmar coal, was operated by Kaiser Resources Ltd, and was imported by Mitsubishi Corporation on behalf of seven major Japanese steel producers. Shortly after the Balmar coal contract was signed a second 15 year contract was signed for the import of Luscar coal for 1 Mt annually. This contract was signed by Mitsubishi and C.Itoh & Co, and again represented the wishes of five major Japanese steel makers, all of whom were partners in the Balmar coal contract. The signing of these two contracts and the others that were to follow in the 1970s, began to transform the whole structure of the Canadian coal mining industry.

Up to that time most coal production had been concentrated in the eastern Atlantic provinces of New Brunswick and Nova Scotia. Unfortunately for the Canadian coal mining industry most of the country's industrial centre is located in the south of Ontario and Quebec. This meant that coal in the mid-western US states and the Appalachians was often much closer in physical location than indigenous (west) Canadian coal, and although some capacity is actively mined in Nova Scotia and New Brunswick this is still not as close as US coal. The result (at the time) was that many of the eastern coal mines were in decline. The new interest by the Japanese firms, and the expansion that was envisaged, led several Canadian firms to contemplate shipping the western coals to the industrial heartland by ship from Lakehead (Lake Superior).

The two Japanese coal contracts mentioned above called for the delivery of the coal from 1970, and it was indeed that year that marked the revival of Canadian exports. Shipments rose from 1.2 Mt in 1969 to 4.0 Mt in 1970.

The emergence of these long-term contracts lead to some very optimistic forecasts of the future for the industry. For example, in 1970 Canada's Department of Energy, Mines and Resources predicted[23] a rise in total Canadian production from 9.7 Mt in 1969 to about 41 Mt in 1975 and 54 Mt in 1980. In the event total production in 1975 was only 25 Mt and only 37 Mt in 1980. Nevertheless the report did accurately assess the principal market for Canadian coal as Japan, but in view of the long term contracts that had been signed and those that were being contemplated, this was a reasonable assumption.

Canadian coal production

Throughout most of the 1960s domestic coal production was steady at

22. Tex Report Ltd *1981 Coking Coal Manual*, (Tokyo, Japan).
23. Northern Miner: "Canadian Coal", No 39, December 1970.

around 9 Mt annually. However this included a large proportion of sub-bituminous coal and lignite. Domestic bituminous coal consumption is still relatively minor although there has been some increase since about 1979, and the second oil price rise increase. Much of Canada's hard coal requirements are fulfilled by imports primarily from the USA, although some other countries, for example Poland, have shipped small quantities from time to time.

In anticipation of the requirements of Japanese steel mill contracts, domestic bituminous coal production rose by 63% in 1970 compared to the previous year. This is further confirmed by noting the location of the production, which was in the two western provinces of British Columbia (Balmar coal) and Alberta (Luscar coal).

Despite the security of income that these early contracts provided the Canadian coal industry was well aware of the danger of too heavy reliance on one major customer, Japan. To this end Canada's Deputy Minister of Energy, Mines and Resources called attention to the "danger" as early as 1971 in a speech to a Canadian conference on coal[24] when he stressed the need for more diversification of markets for West Canadian coal in order to ensure long-term economic stability for the coal mining industry. Within two months the Federal Government echoed this view[25] and called for more vigorous planning and development aimed at prompting long-term stability in order to secure "the orderly growth and optimum development of this valuable resource".

However other overseas markets were not easily forthcoming at that time for the Canadian export programme, which had been built on the back of the world coking coal shortage at the end of the 1960s, faced a downturn in steel production in many countries. Relatively small tonnages were consistently shipped to the European market; in particular the main coal users: France, West Germany, Italy and Denmark but until 1975 there was little penetration in to the Pacific market. This market, which simply by the location of the Canadian mines marked the "natural" outlet for West Coast coal had, until that time few significant steel producers other than Japan. Not until South Korea's steel industry (with heavy government backing) started to develop did Canada have an alternative major market. It was natural therefore that Canadian companies should be among the first to penetrate this market. Shortly after this, contracts were signed with Mexico and Brazil, and eventually with Chile and Argentina.

Nevertheless, despite this broadening of outlets it was still principally the Japanese investors who gave the financial backing that continued to be the main incentive to opening up and developing the new enormously

24. 23rd Canadian Conference on Coal; "Coal Markets", in International Coal Trade, US Bureau of Mines, Oct 1971.
25. Mining Journal: "Outlook for Coal", 12/11/71 in ICT Dec 1971.

Table 21 Canada: bituminous coal production by province (million tonnes)

Year	N.Scotia	N.Brunswick	Alberta	B.Columbia	Canada
1970	1.9	0.4	2.6	3.2	8.0
1973	1.1	0.4	3.8	7.0	12.3
1975	1.7	0.4	4.1	9.6	15.8
1979	2.0	0.3	5.0	9.6	16.9
1980	2.5	0.4	6.2	9.2	18.3
1985	2.8	0.6	7.8	23.1	34.3
1990	3.4	0.5	9.1	24.6	37.6

Source: Monthly "Coal and Coke Statistics" for Canada (Ottawa) various years.

Table 22 Canada: coal production, exports, derived consumption, imports (million tonnes)

Year	Bituminous Production	(−) Exports	= Domestic Supply	+ TOTAL* Imports	Derived Demand
1970	8.0	4.0	4.0	17.1	21.1
1973	12.3	10.9	1.4	14.9	16.3
1975	15.8	11.7	4.1	15.2	18.5
1979	16.9	13.9	3.0	18.0	21.8
1980	18.3	14.3	4.0	16.1	20.1
1985	34.3	27.3	7.0	14.6	21.6
1990	37.7	31.0	6.7	14.2	22.9

*Includes Anthracite
Sources: Statistics Canada Industry Division, various years.

expensive mine and transport systems. Of all the infrastructure require-ments that were needed, it was perhaps the unit train concept that did most to make the shipment of Canadian coals to Japan economically viable. The distances involved from mine to the port (frequently around 800–1,000 miles), and the terrain difficulties that had to be overcome meant that the freight cost on this leg was critical to the whole operation. The concept of a unit train, often two miles long, which shuttles back and forth without actually stopping, was an innovation that, had it not succeeded, would have put the whole project in grave doubt.

Canadian coal consumption by sector

Although the Canadian steel industry has grown over the last decade there has been relatively little change in the quantity of coal consumed by the industry. The explanation lies in the technical changes that have occurred in the industry; for example there has been a strong growth in production of electric arc steel making, which has increased from 17% of total production in 1972 to a peak of 27% in 1981 and is currently running at about 25%. Also there has been a substantial increase in the continuous casting ratio from around 12% in 1975–1976 to 76% today.[26]

Table 23 Canada: primary energy consumption (MTOE)

Year	Coal	Oil	Gas	Hydro	Nuclear	Total
1965	18.1	55.0	22.4	30.7	—	126.2
1970	16.9	73.0	32.7	40.1	0.0	162.7
1973	15.6	83.7	41.8	46.1	3.7	190.0
1975	15.5	83.1	43.1	50.3	3.0	195.0
1979	18.2	90.1	50.1	54.8	8.6	221.8
1980	22.6	87.6	49.3	57.4	9.2	226.1
1985	29.3	68.5	44.8	65.8	14.4	222.8
1990	28.9	74.8	55.0	63.7	16.5	238.9

*For 1965 only, "coal" includes all types of "solid fuel". From 1970 onwards "coal" is only bituminous coal, anthracite and lignite/brown coal.
Source: BP Statistics of World Energy, 1982, 1980, 1991.

Table 24 Canada: primary energy consumption (%)

Year	Coal	Oil	Gas	Hydro	Nuclear	Total
1969	14.3	43.6	17.7	24.3	—	100.0%
1970	10.4	44.9	20.1	24.6	0.0	100.0%
1973	8.2	43.8	21.9	24.1	1.9	100.0%
1975	7.9	42.6	22.1	25.8	1.5	100.0%
1979	8.2	40.6	22.6	24.7	3.9	100.0%
1980	10.0	38.7	21.8	25.4	4.1	100.0%
1985	13.2	30.7	20.1	29.5	6.5	100.0%
1990	12.1	31.3	23.0	26.7	6.9	100.0%

Table 25 Canada: hard coal consumption by sector (million tonnes)

Year	Electricity	Coking	Other	Total
1970	8.2	7.3	5.3	20.8
1973	7.7	7.7	2.2	17.6
1975	7.7	7.3	2.8	17.8
1979	10.9	7.9	1.7	20.5
1980	12.5	7.3	1.6	21.4
1985	13.2	6.3	2.0	21.5
1989	14.6	5.9	1.9	22.4

Sources: 1970–1989.[27]

26. IISI, *Steel Statistical Yearbook 1990* (Brussels 1990).
27. OECD *Energy Statistics 1970–1985 and 1980–1989* (IEA, Paris).

POLAND

At the beginning of the 1980s the uncertain political situation in Poland: unrest amongst the workers, strikes, riots, and demonstrations, threw the once stable coal production process into an entirely new phase with serious questions raised about the country's ability to expand further—or even maintain existing output levels. Until 1979—despite several years of rumbling discontent amongst the labour force—the coal industry in Poland was in general terms a model of growth and achievement. From 1984 to 1988 the country returned to close to its 1979 levels of output and exports. Since 1989 a sharp decline in production and exports has been seen.

Reserves

At the end of 1988 hard coal reserves (proved amount in place) were estimated[28] at 65.9 billion tonnes. Proved recoverable reserves were at the same time estimated at 29.7 billion tonnes.

The main existing area of interest within the country is the Upper Silesian basin located in the south of the country. Reserves in this field are estimated at some 54.3 billion tonnes of which 32% are[28a] classified as coking coal. This particular area is extensively developed and extends over some 4,500 square kilometres. The area is a traditional mining area within the country with active exploitation having existed for over 2,000 years. Regarding the quality of the coal: there is a wide variation with calorific value ranging from 4,500 to 7,500 cal/g, ash content from a few per cent up to 30% and total sulphur varies from less than 1% up to 2%.

The two other areas of interest in the country are the Lower Silesian field and the Lublin Basin. The Lower Silesian field is located in the south western part of the country close to the border with Czechoslovakia. Reserves in this field at 400 million tonnes are only one hundredth of the upper Silesian fields. The coking coal product is used mainly as a blending coal for use with Upper Silesian coal.

The Lublin Basin is situated in the south western part of the country and, relative to the other two areas, is a new development within the country. It is currently undergoing its first phase of industrial exploitation, and although the first two mines are under construction future development depends upon the demand for coal in the domestic and export market. On this basis a question mark exists about the economics of future development. Reserves in the Lublin Basin are estimated at 6.8 billion

28. "Surveys of Energy Resources 1989", World Energy Conference, London.
28a. (1) OECD: "Energy Statistics" 1970–85 and 1980–1989 IEA, Paris; (2) "Surveys of Energy Resources 1989", World Energy Conference, London.

tonnes of which 1.1 billion is "type 34 gas-coking coal" and the remainder is steam coal.

Poland: primary energy consumption

Solid fuels form the most important contribution to Poland's energy supply. Since 1955 coal has never accounted for less than 80% of total primary input. Liquid fuel consumption was increasing up to 1977, but since 1979—and the second OPEC oil price rise—consumption has been on a declining trend.

The reason why the switch to oil did not parallel the switch to coal that was seen in the West in the 1960s, was that in general coal deposits were more easily available than other sources of energy, and because Poland was to a large extent dependent upon crude oil supplies from the Soviet Union. Crude oil began to flow through the Druzbha pipeline from the USSR at the end of 1963. In the previous five years oil imports from the Soviet Union had increased from 4 million tonnes in 1958 to 9 million tonnes in 1962.

So it can be seen that there was a substantial move towards oil—although in absolute terms this was not matched by the growth in coal consumption. In fact for the period from 1960–1970 liquid fuel consumption trebled (from 3.7 MTCE), a much faster growth rate than for coal where consumption grew by 36% over the period. Nevertheless, in terms of market share, coal was on a constantly declining trend falling from 97% in 1955 to 80% in 1979. Natural gas consumption exhibited the strongest growth in this early post-war period and from 1960–1970 showed an eightfold increase.

Table 26 Poland: apparent primary energy consumption[29] (MTCE)[29a]

Year	Solid	Fuels*	Liquid	Fuels	Nat.Gas	Hydro	Total
1955	69.1	96.6%	1.5	2.1%	0.7 1.0%	0.2 0.3%	71.5
1960	87.6	94.8%	3.7	4.0%	0.9 1.0%	0.2 0.2%	92.41
1965	102.2	91.9%	7.1	6.4%	1.8 1.6%	0.1 0.1%	111.2
1970	119.6	86.1%	11.8	8.5%	7.3 5.3%	0.2 0.1%	138.9
1975	141.8	83.5%	17.9	10.5%	9.8 5.8%	0.3 0.2%	169.8
1980	144.3	81.8%	20.8	11.7%	12.4 7.0%	0.4 0.2%	177.9
1985	142.3	82.5%	17.5	10.2%	12.4 7.2%	0.2 0.1%	172.4
1988	144.7	80.4%	20.3	11.3%	13.8 7.7%	1.1 0.6%	179.8

* Solid fuels includes hard coal and brown coal.

29. UN *Energy Statistics Yearbook*, Various years.
29a. Million tonnes coal equivalent.

Poland: coal production

It has been claimed[30] that 1958 marked a turning point in the productivity of labour that had been declining from 1951–1956. This was largely as a result of the effects of increased capital investment in the post-war reconstruction period. The effect of this increased productivity over the two years 1958–1959 was an increase in output per man of over 100 kg per day. Increased efficiency resulting from technological and other improvements led to a fall in coal prices and was therefore a further incentive encouraging coal use.

The objective at that time (1957–1958) was to obtain a hard coal production of about 103 million tonnes in 1960, and about 112–113 million tonnes in 1965. In the event the actual figures turned out at 104.4 million tonnes in 1960 and 118.8 million tonnes in 1965.[31] This outcome, where results exceeded expectations, largely reflects the healthy growth and massive capital investment that was taking place in the Polish coal industry at that time. However in addition to this most obvious and measurable improvement other changes were taking place that enabled the industry to make advances. Not least of these was the opening of new mines and a reorganisation of the labour force. In some ways this new interest in coal was in line with the actions being seen in other European countries in that production was directed towards:

"A return to the working order of production capacity and also to make up for pre-War underinvestment. The reorganisation and nationalisation of the coal industries in France, the United Kingdom, Poland and other countries in Eastern Europe by centralising investment decisions, and the putting into operation of them, facilitated these tasks."[32]

To quantify the capital investment that was being undertaken at that time it was estimated[33] that 2,174 million Zlotys were invested in the mines between 1945 and 1949 compared with 11,604 million Zlotys in the following six years and 18,347 million Zlotys from 1956 to 1960—all measured in constant prices.

The two periods 1950–1955 and 1956–1960 also marked a fundamental redirection of Polish investment. In the first period the principal objective was the maximising of output from existing mines, whereas in the second half of the decade, investment was directed towards opening new pits. The 1956–1960 "Five Year Plan" called for the construction of six new coal mines to be completed between 1959 and 1965 and work to start

30. Economic Commission for Europe, *The Coal Situation and and Prospects in Europe in 1958/59.* (UN, Geneva, 1959).

31. *Ibid.*

32. Economic Commission for Europe *The Coal Situation in Europe in 1962–63.* (UN, Geneva 1963).

33. *Ibid.*

construction of seven new mines of which five were designed for coking coal necessitating relatively high investments.

Despite this massive investment and the rapid expansion of coal production (+40% from 1950 to 1962) this growth rate barely kept pace with the growth in industrial activity in the country, which over the same period rose by over 270%. In fact the demands of the industrial and domestic sectors resulted in a cut in exports in the mid-1950s.[34] In 1955, 24 million tonnes of hard coal was exported, but despite growth in world demand this fell to 19 million tonnes in 1956 and 14 million tonnes in 1957.

Table 27 Poland: hard coal production and exports by type (million tonnes)

Year	Production	Exports	Coking	Thermal	Total	%Exports
1945	27.4		—	5.6	5.6	20.4%
1950	78.0		—	26.5	26.5	34.0%
1955	94.5		—	24.0	24.0	25.4%
1960	104.4		—	17.5	17.5	16.8%
1965	118.8		0.6	20.4	21.0	17.7%
1970	140.1		5.2	23.6	28.8	20.6%
1975	171.6		10.7	27.6	38.3	22.3%
1980	193.1		6.3	24.6	30.9	16.1%
1984	191.6		10.9	32.2	43.1*	22.5%
1985	191.6		10.8	25.4	36.2	18.9%
1990	145.0		10.5	17.5	28.0	40.6%

*Peak year
Source: CHZ "Weglokoks" (Katowice, Poland—various years).

The source for these investments was, and remains, Central Government funds, although the International Monetary Fund and the World Bank have also contributed to the resources available to Central Government from year to year as described in a United Nations report of 1963[35]:

"In Poland investments are financed from the following sources: (a) the central plan budget; (b) bank credits; (c) undertakings amortisation funds; (d) under-takings profits. Capital repairs are financed by that part of the amortisation fund reserved for this purpose. The dividing line between capital repairs and invest-ments is determined by the provisions of the current economic plan."

Undoubtedly the rapid development of Polish coal after the War was due in part to the demand exerted by the Soviet Union which was able to exploit its role as the leading member of the East European bloc. An example of this was the Polish–Soviet coal agreement whereby Poland supplied coal at below market price. However the situation could not

34. *Ibid.*
35. Economic Commission for Europe, *The Coal Situation in Europe in 1961/62*, (UN, Geneva 1963).

continue indefinitely as a drain on the fragile Polish economy and, following disturbances in Poland in 1956, the Soviet Union agreed to a reimbursement as compensation for the low prices.[36]

During the mid-1960s activity was focused upon increasing productivity to satisfy the internal energy demand and the growing export market. Work continued on the opening of new mines and increasing mechanisation. Attention also focused upon the exploitation of brown coal mines where output rose from 6 million tonnes in 1955, to 11 million tonnes in 1962 and 15.5 million tonnes in 1963. In many cases the brown coal extraction was achieved by open-cast methods and therefore the rapid rate of expansion was not unreasonable. In 1963 there was intensive construction of brown coal mines at Kazinierz, Patnow and Turow.[37] Poland was something of an exception among the other Eastern European countries at that time, because in every other member of this group, brown coal production exceeded hard coal output.

During this period of the early 1960s, the rapid growth in industrial activity in the country meant that export coal was being limited by internal demand. It was estimated[38] that between 1955 and 1963 total coal production increased at an annual rate of 3% while industrial output rose at an average of 9.1% pa for the same period. The effect of this was that whilst exports were 24 million tonnes in 1955 by 1960 they had fallen to 17.5 million tonnes and had only slowly increased to 21.0 million tonnes by 1965.

It soon became obvious, due to the growing demand for coal, that these output increases could not be achieved simply by increasing productivity. Extra manpower was essential and a plan of 1967[39] recognised that the number of miners would have to increase by 11,500 or 3.4% in the period 1966–1970 as compared with 1965. Clearly the importance of coal to industry, both as an energy supply source to the Soviet Union and as a major hard currency revenue earner, was being recognised and every effort was being directed towards its exploitation.

Similar major advances were expected of brown coal, principally directed towards the generation of electricity: output was to be increased by 11 Mt to 34 Mt in 1970 and it was estimated[40] that the share of power generated from brown coal would rise from 23% in 1965 to 34% in 1970. In fact the outcome was only marginally off-target with brown coal actually accounting for 32% of electricity production.[41]

36. P.Marer, "The Political Economy of Soviet Relations with Eastern Europe", in S.J.Rosen and J.R.Kurth (eds) *Testing Theories of Economic Imperialism* (Lexington, USA, Lexington Books, 1974).

37. Economic Commission for Europe *The Coal Situation in Europe in 1963/64 and Future Prospects*, (UN, New York, 1965).

38. *Ibid.*

39. Economic Commission for Europe *The Coal Situation in Europe in 1965 and its Prospects* (UN, New York, 1967).

40. *Ibid.*

41. UN *Annual Bulletin of Electric Energy Statistics for Europe—1982* (New York, 1968).

Despite the fact that the majority of Polish coal exports have been thermal coal, with coking coal exports developing only from the mid-1960s as specialist mines were developed, it is interesting to note that in the period before the first oil price rise in 1973, little future was seen for thermal coal exports. A United Nations report in 1968[42] commenting upon some decisions made in 1967 said:

"The East European energy picture will be dominated by solid fuels for many years to come. Increased production in Poland—by far the most important producing country in Eastern Europe will counteract reductions elsewhere. This does not, however, change the fact that in 1967 important decisions were taken which finally, will lead to the energy pattern in Eastern Europe approaching that of Western Europe in the sense that hard coal consumption will depend mainly on the growth of iron and steel production, and that of brown coal on the growth of power production."

The emphasis was clearly on hard coal for coking supplies, for internal consumption in steelworks, and for exports, while power generation was expected to be supplied by brown coal.

Polish coal situation 1970–1990

During the 1970s Polish hard coal production experienced a period of rapid expansion: output rose by 61 million tonnes (from 140 Mt. to 201 Mt.) from 1970 to 1979, a 44% increase. In 1980 and 1981 the country experienced two years of disruption but in the following three years showed strong signs of recovery and exports reached a peak in 1984 (at 43 million tonnes). The energy sector in Poland is rated very highly in terms of importance to the economy; in 1990 it was estimated[43] that "it accounted for about 10 per cent of gross domestic product". The difficult economic situation however means that every investment decision has to have the highest consideration. As the largest foreign hard currency earner hard coal exports are necessary to reduce the international debt. However without further loans either from international organisations such as the International Monetary Fund, or the Soviet Union, it will be difficult to expand the industry further without the always massive investment that is required.

42. Economic Commission for Europe *The Coal Situation in Europe in 1967 and its Prospects* (New York 1968).

43. International Energy Agency *Energy Policies Poland—1990 Survey* (Paris, 1991).

CHAPTER 3

Major consumers—patterns of demand

EUROPE

In Western Europe demand for energy grew rapidly in the post-war period rising from 598 million tonnes of oil equivalent (MTOE) in 1960 to 1,409 MTOE in 1990.[1] Production of primary energy over the same time period increased from 396 MTOE to 822 MTOE. Overall energy production is currently at a peak; however when coal is isolated from the aggregate energy view a different picture emerges. In 1960 demand for coal in Europe was 337 MTOE; by 1989 this had fallen to 257 MTOE. This would not have caused any problems if coal production in Europe had fallen according to demand; the reality however was that coal production declined more rapidly and over the 1990s seems set to decline further.

Table 28 Western Europe: primary energy production/ consumption (MTOE[2])

	Production	(of which Coal)	Consumption	(of which Coal)
1960	396	308	598	337
1970	410	234	1,028	274
1975	492	204	1,118	220
1980	645	203	1,246	254
1985	787	182	1,262	261
1989	796	175	1,321	257

Clearly this has been a period of change in the energy industry and to look back at the energy situation in context might be useful at this point. It has been argued[3] for instance that the two years of peak coal production (1957–1958) represented the turning point in the European coal/energy market.

Historically, the years from the end of the Second World War up to 1957–1958 represented a period of energy shortage with coal production increasing by 1.2% pa, (1953–57) whilst energy demand was increasing by 5.5% pa. Coal was still the main European energy source by the end of the 1950s representing 65% of primary supply, but only ten years earlier, coal had supplied 84% of total energy needs. The availability of large volumes

1. *BP Statistical Review of World Energy* (June 1991).
2. *Energy Balances of OECD Countries 1980–1989* (OECD Paris 1991).
3. *Twenty-Five Years of the Common Market in Coal—1953–1978* (Commission of the European Community, 1977).

of oil from the Middle East with its associated advantageous handling characteristics, combined with a falling real oil price led to a rapid expansion in the use of oil, which more than tripled in use.

This early peak consumption of energy in 1957–1958 was due in part to the particularly severe winter of 1956 causing a rundown in power station stocks and a desperate attempt by the utilities to rebuild their supplies. At the same time the Suez Canal closure led to increased demand for vessels and transatlantic freight rates rose to $15 per tonne for US coal. The result was that US coal was more expensive than domestic supplies, and pressure was put on European mines to increase production.

In 1957, coal stocks were built up again and production was at its peak, but the underlying weak economic situation was about to give way leading to diminished coal demand in 1958. Industrial activity, which had in fact first shown signs of slowing in 1956 declined further throughout 1958, such that the period is now recognised as one of industrial recession. As demand started to fall away so consumers tried to reduce their expensive stocks to more appropriate levels. At the same time the re-opening of the Suez Canal led to a slump in freight rates as shown in Table 29.

Table 29 CIF price US coking fines delivered to Rotterdam

End Year:					
1953	1954	1956	1957	1958	1959
$13.66	$15.94	$26.81	$14.38	$13.52	$13.58

As can be seen from the figures in Table 29, in the course of only two or three years the coal market had turned around. From a superficially healthy position as the predominant energy source unable to keep up with demand it had—through the industrial slump, high stockpiles, high prices, and cheap oil—opened the door for the rapid expansion of oil.

It can be argued that had none of these problems existed in the coal industry, it still would have had to lose market share to oil as a more convenient and cheaper energy source, but these problems still had a significant part to play in the coal market in the 1960s.

Most European coal producers entered the 1960s from a position of weakness: falling demand and intense competition cut consumption by 50 Mt in only five years (1956: 545 Mt, 1960: 494 Mt). In this period European railways alone cut their consumption by 10 Mt per year with the increased electrification of track, whilst industry, switching from coal to oil-fired boilers, cut consumption by 16 Mt.

This rapid switch out of coal towards oil led to political measures being taken to slow down the movement. West Germany, Austria and Belgium all introduced special taxes on imported fuel oil designed to bring the balance between supply and demand nearer to equality.

In 1961, thermal power stations were the major area for increased coal usage, delivery to them increased by 6.4 Mt to 123 Mt or 6% over 1960. This was due partly to the continued increased demand for electricity, partly to the fact that many oil-fired plants had simply not yet been built, and finally to the slow down in hydro-electric generation which in 1960 accounted for more than a third (225.2 bn kWh: total 569.5 bn kWh) of total electricity produced.

Table 30 OECD Europe: coal consumption by sector (million tonnes)

Year	Power	Solid fuel*	Industry	Gas/Rail	Domestic	Other	Total
56	108.6	155.4	95.5	78.1	73.7	33.7	545.0
57	115.0	161.2	91.9	73.4	70.3	30.7	542.5
60	116.6	150.5	83.4	58.6	61.2	23.4	493.7
65	146.4	146.4	71.5	38.7	47.3	15.7	466.0
70	147.9	143.2	38.7	12.4	39.8	12.0	394.0
73	145.6	126.8	25.9	5.3	25.9	7.6	337.1
75	133.6	122.5	22.5	3.3	21.1	6.8	309.8
79	186.3	106.1	20.1	2.0	18.6	7.0	340.1
80	197.4	105.3	20.5	1.6	16.0	5.6	346.4
85	192.5	92.7	30.7	0.5	17.1	5.8	339.3
89	200.2	80.7	31.6	0.0	11.3	4.4	328.2

*Solid fuel = consumption for cokeries and patent fuel.

One sector of coal consumption that was in decline throughout the 1960s was the use of coal for the production of gas. In 1960, this sector consumed some 35 Mt, which had already fallen from 44 Mt in 1956, and was destined to decline to less than 9 Mt by the end of the 1960s. This was due to three factors: first, it was found that the gases produced in the coke ovens could be substituted for coal gas; secondly, the large-scale exploitation of the natural, gas which was just beginning especially in the Netherlands; and finally the growing trend for using petroleum for the production of town gas. In the domestic sector a slow decline in consumption was occurring, for example up to 1955 deliveries to households were about 76 Mt annually, by 1960 however, this had fallen to 70 Mt.

Although demand for coal was declining it was clear, even by 1960 that two sectors would be the main consuming areas. Those two sectors (coke ovens and power stations) remain to this day whilst the railways and gas coal have virtually disappeared. This trend continued more or less uninterrupted throughout the 1960s and early years of the 1970s.

Summary of period 1956–1972

The whole period had been characterised by shrinking markets. Compare for example the situation in 1956 with that in 1970:

Table 31 Shrinking markets—1956 and 1970 (million tonnes)

Consuming Sector	1956	1970
Patent Fuel Plants	20.9	11.5
Coke Ovens	134.5	131.7
Gas Works	44.2	8.7
Power Stations	108.6	146.4
Coal Mines	19.7	5.3
Railways	33.9	3.5
Iron & Steel Industry	11.1	5.5
Other Industries	73.7	37.4
Other Sectors (Domestic)	98.4	45.9
Total	545.0	395.9

In this period the effective market shrank to just two sectors: coke ovens and power stations. All other sectors, with the possible exception of patent fuel plants, suffered from the inconveniences of coal. This was especially true in those sectors where its use was already marginal such as the domestic sector, when it was unable to compete with the modern clean image of gas or electric central heating. Technical advances in the coke ovens and power stations even reduced consumption in the two largest markets. The electrification of the railways and the change from coal gas to natural gas, were both circumstances that were inevitable evolutions to more efficient systems given the availability of cheap electricity resulting from cheap oil, and plentiful supplies of natural gas from indigenous sources.

Demand for coal in the European Community after the 1973 oil price rise

In the European Community expansion of the economy, as measured by GDP growth and industrial production, progressed at a rapid rate in 1973 with 5.6% GDP growth achieved for the Community as a whole. Inflation continued to be a serious problem and ranged between 6% and 11% for individual countries.

The most marked feature of the year was the oil price rise imposed by OPEC countries. The effect on economic activity commenced during the last quarter of the year when a slowing down of energy consumption was observed in those sectors most acutely affected by the price rise. Balance of payments were adversely affected for all European countries as all were net oil importers. Uncertainty over oil price policies did not assist investment by businesses and many expansion programmes were suspended or postponed. As a consequence of this fear of loss of supplies, an immediate interest was shown in alternative fuels. As has been mentioned the transition from coal was in many cases a one way process (by the railways and the domestic sector), and that even sharp increases in the oil price structure could not have an equally short-term and favourable effect on the use of coal.

General energy situation

Total inland consumption of energy declined in 1974 and 1975 following the OPEC oil price rise. Consumption recovered in 1976 but only to the level seen in 1973. The breakdown by fuel used is shown in Table 32.

Table 32 European Community: primary energy consumption (MTOE)

	1973	1974	1975	1976
Coal	196.0	189.8	167.1	178.4
Lignite	26.1	27.4	26.7	29.2
Oil	554.8	519.2	476.4	506.9
Natural Gas	117.9	134.4	142.2	153.1
Prime Electric/Other	41.2	46.3	52.1	48.5
TOTAL	936.0	916.8	864.5	916.1

It can be seen from Table 32 that the recession acted adversely against coal in the short term. Only two energy resources were "unaffected" by the oil price rise: natural gas, which was largely of indigenous origin, and primary electricity, which is largely nuclear and hydro-electricity. It should be noted however, that the recession was only partly the cause of the cutback in energy demand in 1975; the mild climatic conditions and the introduction of energy conservation measures were also having an affect.

Demand for coal

Although coal demand fell, in line with the industrial and economic recession, it was observed that coal was not able to act as a substitute for oil in the short term.[4] This was also seen after the closure of the Suez Canal in 1956 when supply could not keep pace with demand. After the 1973 oil crisis, the coal industry again could not keep pace with the loss of oil supplies.

The European coal market has, as one of its characteristics, a lack of flexibility and ability to respond rapidly to changed market conditions. Adjustments have been made to maintain the energy balance by means of stock drawn down or increased imports.

4. P. Rogers, "Developments in Seaborne Coal Trade 1960–85", (PhD Thesis 1988).

(i) Power stations

Table 33 Hard coal consumption in thermal power stations (million tonnes)

Power Station	1971	1972	1973	1974	1975	1976
Consumption	121.4	106.6	118.8	102.9	100.6	120.9

In 1972, the market share of coal in the production of electricity in all coal producing countries of the Community (except for the United Kingdom) decreased in tonnage and percentage terms. In France, positive steps had been taken to encourage the use of oil and several coal-fired power stations were converted to oil. This declining trend of coal consumption was foreseen at the time as one likely to continue.[5] In Belgium for example, no plans had been established to increase domestic coal production, additional energy requirements would be covered by oil. In the Netherlands, a similar situation existed for both coal and oil, with the exception that additional energy requirements would be provided by natural gas.

In the Community (of nine) in 1971, indigenous coal supplies provided 90% of the 121.4 million tonnes (Mt) of thermal coal consumed at the power stations, ie 12.4 Mt had to be imported.[6] By 1972, with the fall in demand (to only 106.6 Mt consumed), imports fell to 9.5 Mt.[7] The main suppliers of this coal were Poland (4.2 Mt) and the USA (3.5 Mt). The two largest coal producers in the Community—West Germany and the United Kingdom— were also the two main importers although at 2.6 Mt and 3.1 Mt respectively, the tonnages accounted for only 7% and 5.8% of coal consumption in the power plants of the country in question.[8]

By 1973, net electricity production had reached its highest level at 987 GWH with hard coal accounting for 30% of generation and with oil (ie petroleum products) achieving their highest ever level at 32%, a level that has not been achieved since. Table 71 shows the development for the period 1975–1984 of EC coal supplies to the power stations by source.

As mentioned above, by 1972 several countries had converted from coal to oil firing. The United Kingdom in the period 1971–1974 was the only country to put "coal only" stations into operation. In fact the UK had the largest proportion of coal fired generation—at 70% of thermal capacity— against 40% for the EEC as a whole.

5. *Official Journal of the European Communities*: 16/4/73, Vol 16, No C23, "General Coal Market Situation—Forecasts for 1973".
6. *Ibid*. p 8.
7. *Official Journal of the European Communities*: "General Coal Market Situation—Forecasts for 1974" 15/11/74, Vol 17, No C141, p 17.
8. *Official Journal* C147, p 17.

The year 1975 was notable as it was the first time in 25 years[9] that a decline in electricity production was seen in some member countries. There were two main reasons for this decrease: the recession and energy conservation measures. Although it is not possible to quantify the proportion due to each element, it is believed that the recession had by far the greater effect as conservation measures at the time were not very advanced.

It might also have been thought that two years after the quadrupling of oil prices some advance might have been made in switching towards coal-fired generation, but the figures for generation and consumption for 1974 and 1975 are nearly identical and show that for the Community as a whole no clear trend had been established. In fact it can be argued that the situation had positively deteriorated as the coal strike in the United Kingdom in 1974 prevented the desired amount of coal from being utilised—in other words had there not been a strike, the figures for the Community in 1975 would have indicated a further sharp fall in consumption rather than staying roughly level.

In 1975, the United Kingdom increased its power station hard coal consumption by 5.5 million tonnes and, as the total Community consumption for coal actually declined by 2.3 million tonnes, this implied a net switch away from coal in 1975 in the other member countries of 7.8 million tonnes. The explanation for this switch is entirely dependent upon developments in West Germany.

Table 34 European Community: power station consumption of coal (thousand tonnes)

Country	1974	1975	Change
W.Germany	34,339	26,345	− 7,994
United Kingdom	53,984	59,457	+ 5,473
Others	14,564	14,799	+ 235
Total	102,887	100,601	− 2,286

Sources:[10]

In West Germany in 1975, coal consumption was reduced despite legislation that had been introduced aimed at maintaining the average annual coal burn at 30–33 MTCE. In that year, coal consumption fell to 26.3 MTCE (from 34.3 MTCE in 1974) as two new nuclear and two new gas-fired plants were commissioned.

9. *Official Journal of the European Communities*: 28/6/76, Vol 19, NoC146, p 4, "General Coal Market Situation 1975 and Forecasts for 1976".
10. *Ibid. Official Journal* C146, pp 17 and 18 and *Official Journal* C156, pp 19 and 20 (4/7/1977).

These new plants alone accounted for 4,163 MW of capacity, roughly equivalent to 7.3 million tonnes of coal. Additionally, electricity generation was on average lower than for the Community as a whole, and the lignite burning stations increased their market share.

By 1976, economic recovery, as shown earlier by the measure of GDP and industrial production, was showing strong signs of growth, and the previous declines in electricity production were all reversed such that every Community country showed an increase. The summer of 1976 was an exceptionally dry period in Europe and this severely affected the capabilities of the hydro-electric plants where generation dropped from 130 GWH in 1975 to 114 GWH in 1976, thereby putting extra pressure on thermal generating plants as the marginal supplier. Nuclear capacity was expanding at this time, but it has in general been used as a base load source with thermal plants supplementing it.

The shortfall in hydraulic capacity was, therefore, satisfied by extra coal demand which increased by 20 million tonnes in a year. This effect was particularly noticeable in France where the government had imposed an upper limit on oil consumption, thereby leaving coal as the only practical alternative. In France in 1976, this measure had the effect of increasing the power stations consumption by 60% to 15 MTCE from 9.5 MTCE. Similar government intervention in Belgium and Germany has been estimated[11] to have increased power station coal burns by 25% in 1976 although some of this increase is inevitably due to the increase in industrial activity and the breakdown of a nuclear plant in West Germany. In Belgium the government intervention took the form of subsidies to the coal industry enabling low grade coking coal to be sold to the power stations.

In the space of four years (1973–1976), government attitudes towards coal had changed completely from one of positively discouraging coal to one of promoting coal. Even with this change in attitude however, it was soon discovered that the coal industries of the Community's producer countries were, in general, unable to respond to changed market conditions. Energy supply was recognised to be a long-term problem that could not be solved by legislation even if that legislation was the "optimum" to encourage coal use. The inflexibility of the coal industry at home meant that to achieve government objectives additional coal requirements had to be covered by extra imports.

(ii) Steel industry

Demand for metallurgical coal is highly correlated with pig iron production as coke is required in the blast furnace for its manufacture.

---

11. *Official Journal of the European Communities*: 4/7/77, Vol 20, No C156, "The Community Coal Market in 1976 and Forecasts for 1977".

Demand for pig iron is in turn closely related to crude steel production levels and the fortunes of the steel industry in particular. Table 35 shows in detail the relevant developments over the period 1970–1990.

The year 1972 was the first that crude steel production reached 150 million tonnes; and reflected the strong growth seen in the economy, as mentioned previously. Pig iron production exceeded 100 million tonnes for the first time to reach 107 million tonnes.

Table 35 European Community: steel/pig iron production (million tonnes)

Year	Steel	Continuously Cast%	Pig Iron
1970	138	na	99
1973	151	9%	107
1975	126	17%	88
1979	141	31%	98
1980	142	39%	89
1985	105	71%	85
1990	110	89%	92

Source:[12]

The effect of the oil price increase was to slow down demand for steel and consequently the rate of increase was reduced in 1974, although this still meant a new record level of production of 155.6 million tonnes. The long-term effects of the recession were not foreseen and the European Commission, as an official comment, stated:

"The energy crisis should not stop the general tendency for the iron and steel industry to expand."[13]

Whilst this comment may have been referring to 1974 alone, the implication was that the industry would expand further.

In 1974, the coke ratio (quantity of coke required to produce one tonne of pig iron) stood at 537 kg. This requirement had been falling steadily for the Community of Six since 1957 when 972 kg were needed. This was achieved largely by the injection of fuel oil in the blast furnace in place of coke. This was desirable for two reasons:

1. Fuel oil was (by 1972) a cheaper fuel than coke, and therefore the blended mix of oil and coke was cheaper to the producer.

2. The injection of fuel oil was found to be technically superior in that it enabled better combustion in the furnace as it was more controllable, ie it could be regulated better and its effect was more immediate than coke.

12. IISI *Steel Statistical Yearbook 1980/1986/1990* (Brussels, Belgium).
13. *Official Journal of the European Communities*: Vol 17, C141, "General Coal Market Situation—Forecasts for 1974".

There was, however, a limit to the degree of substitution that was possible, but this had certainly not been reached by 1974 as the rate was to dip under 500 kg/Tonne in 1978 (498 kg/Tonne).

The supply difficulties encountered with fuel oil at the end of 1973 and the ready supplies of coke meant that some blast furnaces substituted coke for this oil. The situation varied from country to country however. Italy, for example, showed no change, whilst Belgium and West Germany increased their coke input. The decisive factor regarding this decision to reduce fuel oil injection or not, depended quite simply on whether alternative supplies of coke were available.

By 1975 the steel industry was feeling the full effect of the oil-based recession, and crude steel production fell by 30 million tonnes to 125 million tonnes. This cut back had "knock-on" effects for pig iron and coke consumption, which fell by nearly 20 million tonnes to 66 million tonnes. On average, the cutback for the Community was around 20%.

By 1976, some recovery was seen in crude steel production but by the end of the year this turned out to be weaker than originally hoped for.[14] This difference between forecasts and out-turn was assumed at the time to be due to two reasons:

"In the first place, the revival in consumer demand in 1976 did not lead to a comparable rise in investment in capital goods. Secondly, the Community's exports of steel and capital goods fared less well than those of some other countries, notably Japan."[15]

The steel producers were severely affected by the oil price rise. The increase had two principal effects: first, it cut consumer demand for steel products as, at the corporate level, a greater proportion of revenue was taken by energy costs, while at the individual level net disposable income decreased as again a higher proportion of income was spent on fuel costs. Secondly, it initially increased the coke rate and thereby the demand for coke by the reduction of oil injection, but it was found that a relatively small reduction in the fuel oil price was enough to encourage its injection, again due to the technical advances that had been made possible.

(iii) Other coal consuming sectors

The markets covered in this sector are: gas works, domestic, transport and other industries. As mentioned earlier this sector had been in decline since the mid-1960s. For many industries the move away from coal was in a one way and long-term process (eg the railways and the domestic sector).

14. *Official Journal of the European Communities*: Vol 14, No C146, 28/6/76, "General Coal Market Situation 1975 and Forecasts for 1976".
    15. *Ibid.*

Because of this process, which was coming to an end by the early 1970s, the effect of the oil price rise did not so directly affect consumption.

There was however, an indirect effect caused by the recession and the downturn in consumer demand. The effect was to reduce consumption in the almost miscellaneous category "Other Industry" where demand fell by 25% from 1974 to 1975. As might be expected, a similar cutback was seen in consumption by the iron and steel industry where the coal is used as an energy source and not for the coking process. This sector suffered a 34% cutback but on a much smaller tonnage basis.

JAPAN

Primary energy consumption

Primary energy consumption in Japan grew at an average annual rate of 1.7% per year through the period 1970 to 1984. Since that time (1985–1990) primary consumption has grown at 2.6% pa. That uneven growth disguises some sharp swings: from + 12% in 1973, to − 5% in 1975. These substantial changes to the direction of growth were a direct consequence of the prevailing conditions in the international energy market because Japan is very dependent upon external sources for its energy supplies. In fact since 1970, Japan has never been less than 80% dependent on imported energy. The market share of this imported energy has changed as well, with gas and nuclear energy substantially increasing their proportion. In 1970 for instance, natural gas and nuclear energy provided just 2% of Japan's primary energy requirements, but by 1990 these two categories provided over 21% of total consumption.

Oil, which throughout the period has been the largest primary energy source, reached consumption peaks in 1973 and 1979 when 269 million tonnes of oil equivalent (MTOE), and 265 MTOE respectively, were consumed. Since 1979, the effect of a rapid expansion in nuclear generating capacity, and to a lesser extent gas, increasingly reduced the demand for oil in thermo-electric plant. Oil use for electricity generation peaked in 1979 when 49 million tonnes (Mt) of residual fuel oil was consumed in this way. By 1989 this total was down to just 30 Mt.

The oil price rises of 1973 and 1979 were without doubt the principal reasons for this switch away from oil. However the price rises also brought home to the Japanese people their vulnerability to circumstances beyond their control as far as energy supplies were concerned. This was a situation that they were not unaware of, however, as they were investing heavily in Canadian and Australian coking coal mines in the late 1960s and early 1970s. The oil price rises confirmed this vulnerability and the need to diversify supply sources abroad. In particular, the priority to make more efficient use of raw material imports at home became even more urgent.

Table 36 Japan: primary energy consumption (by fuel MTOE)

Year	Oil	Coal	Gas	Nuclear	Hydro	Total
1970	199.1	60.2	3.6	1.2	21.0	285.1
1973	269.1	53.7	5.3	2.3	17.3	347.7
1975	244.0	54.4	7.7	5.3	19.1	330.5
1979	265.1	50.4	20.3	14.9	19.2	369.9
1980	237.7	57.6	23.4	20.1	20.8	359.6
1985	206.3	73.7	35.9	37.8	20.6	374.4
1990	245.0	75.0	45.4	48.9	21.4	435.7

Table 37 Japan: market share primary energy consumption (%)

Year	Oil	Coal	Gas	Nuclear	Hydro	Total
1970	69.8%	21.1%	1.3%	0.4%	7.4%	100%
1973	77.4%	15.4%	1.5%	0.7%	5.0%	100%
1975	73.8%	16.5%	2.3%	1.6%	5.8%	100%
1979	71.7%	13.6%	5.5%	4.0%	5.2%	100%
1980	66.1%	16.0%	6.5%	5.6%	5.8%	100%
1985	55.1%	19.7%	9.6%	10.1%	5.5%	100%
1990	56.2%	17.2%	10.4%	11.2%	4.9%	100%

Table 38 Japan: primary energy production (MTOE)

Year	Oil	Coal	Gas	Nuclear	Hydro	Total
1970	0.8	27.0	2.9	1.2	21.0	52.9
1973	0.7	16.5	2.6	2.3	17.3	39.4
1975	0.6	12.3	2.2	5.3	19.1	39.5
1979	0.5	11.7	2.2	14.9	19.2	48.5
1980	0.4	11.9	2.0	20.1	20.8	55.2
1985	0.5	10.9	2.0	37.8	20.6	71.8
1990	0.7	5.7	1.5	48.9	21.4	78.2

Table 39 Japan: market share primary energy production

Year	Oil	Coal	Gas	Nuclear	Hydro	Total
1970	1.5%	51.0%	5.5%	2.3%	39.7%	100%
1973	1.8%	41.9%	6.6%	5.8%	43.9%	100%
1975	1.5%	31.1%	5.6%	13.4%	48.4%	100%
1979	1.0%	24.1%	4.5%	30.7%	39.6%	100%
1980	0.7%	21.6%	3.6%	36.4%	37.7%	100%
1985	0.7%	15.2%	2.8%	52.6%	28.7%	100%
1990	0.9%	7.3%	1.9%	62.5%	27.4%	100%

Table 40 Japan: primary energy import dependence* (MTOE)

Year	Oil	Coal	Gas	Total
1970	198.3 99.6%	33.2 55.1%	0.7 19.4%	232.2 81.4%
1973	268.4 99.7%	37.2 69.3%	2.7 50.9%	308.3 88.7%
1975	243.4 99.8%	42.1 77.4%	5.5 71.4%	291.0 88.0%
1979	264.6 99.8%	38.7 76.8%	18.1 89.2%	321.4 86.9%
1980	237.3 99.8%	45.7 79.3%	21.4 91.5%	304.4 84.6%
1985	205.8 99.8%	62.8 85.2%	33.9 94.4%	299.6 80.7%
1990	244.3 99.7%	69.3 92.4%	43.9 96.7%	357.5 82.1%

* (Dependence here defined as "Consumption–Production").
NB: Percentage represents the proportion of that fuel type that is imported—hence nuclear and hydro are zero and not shown.

Japan steel industry developments 1970–1990

The industrial expansion in Japan after the Second World War was extremely rapid as the following figures for crude steel production illustrate:

1955:   9.4 Mt.   1960: 22.1 Mt.   1965:   41.1 Mt.
1970: 93.3 Mt.   1975: 102.3 Mt.   1980: 111.4 Mt.
1985: 105.2 Mt.   1990: 110.3 Mt.

Peak production was actually achieved in 1973 and 1974 when 119 Mt and 117 Mt respectively were produced. In the period up to the early 1960s most of this steel was being used for reconstruction of Japanese industry; exports of steel in 1961 for example reached a new "high" of just 3 million tonnes. Trade in steel expanded rapidly over the last 20 years such that peak exports reached 33 million tonnes. Since 1985 the increase in the strength of the Japanese yen has meant that steel products exports have become increasingly expensive and currently (1991) exports have slumped to around 15 million tonnes.

Table 41 Japan: steel production by process (thousand tonnes)

Year	Open	Hearth %	Oxygen %	Electric Arc	Total
1970	3855	4.1%	73,847 79.1%	15,620 16.7%	93,322
1973	1849	1.5%	96,057 80.5%	21,416 17.0%	119,322
1975	1103	1.1%	84,428 82.5%	16,782 16.4%	102,313
1979	—	—	85,370 76.4%	26,377 23.6%	111,748
1980	—	—	84,130 75.5%	27,245 24.5%	111,395
1985	—	—	74,776 71.0%	30,505 29.0%	105,281
1990	—	—	75,641 64.6%	34,691 31.4%	110,332

Source: Japan Iron and Steel Federation, Various years.

The reconstruction and expansion of the Japanese steel industry was achieved by a determination to obtain the latest technological advances, and this in turn, was achieved by importing "production equipment, processes and technologies from the advanced steel making countries", particularly West Germany and other European producers. This determination carried through to make the Japanese steel industry amongst the most efficient in the world today.

CHAPTER 4

Major exporters

AUSTRALIA

Introduction

Historically, the Australian coal mining industry, and the associated export of that coal, has been established for much longer than most people realise. The records of the New South Wales *Statistical Registers* show that the average tonnage exported for each of the five years 1861–1865 was 138,000 tonnes. After gradual expansion during the second half of the nineteenth century exports first exceeded 2 million tonnes per year for the period 1906–1910.

Much of the growth in the early days of the industry was directly related to growth in the United Kingdom coal export market and specifically the pattern of trade of the typical British tramp. A study of this linkage was made in 1960 by Burley[1] who found that productive coal capacity greatly exceeded indigenous Australian demand and that its price advantage on the export market was largely the product of cheap coal freights.

In those early days (1850 to the First World War) British shipping was dominant in the carriage of international trade. However, the balance of the commodities was such that exports from Europe tended to be manufactured goods, whereas imports tended to be bulky cargoes. The capacity problems that this caused were conveniently solved by the development of the coal industry from the United Kingdom to the extent that the "coals out/grain back" trade from-and-to the UK became a staple part of the routine of many shipowners. The associated growth in world trade that occurred at this time, linked with the industrial expansion that was encouraged by the available supplies of coal from Britain, in turn led to a surplus of cargoes from industrial Europe to the markets of the Far East and West Coast of North, Central and South America. The clear reason for this was the preference of shipowners returning to Europe to route their vessels via South America. This was done because of the prevailing trade winds blowing from West to East. Therefore a cargo that could be picked up in Australia and shipped, to say, San Francisco or Chile, would be a bonus for a shipowner who had discharged in Australia and was then looking to ballast to the East Coast of America for a cargo back to Europe.

1. K.H. Burley "The Overseas Trade in New South Wales Coal and the British Shipping Industry, 1860–1914" in *The Economic Record* (August 1960).

For several reasons this trade did not last: not least was the gradual
replacement of the sailing vessels with the motor ships so that the owner
could now return to Europe via the Atlantic rather than the Pacific. In
addition the break caused by the First World War provided an actual
cessation of shipments. By the time the war was over the Australian
manufacturing industry had advanced to such an extent (relative to the
devastated economies in Europe), that there was no longer surplus coal
supplies available for export. In addition, the development of the Australian wool and wheat trade meant that vessels could now be fixed direct
to Europe and there was little incentive (except at high freight rates) to
seek an intermediate coal cargo.

The Australian coal export industry faced with these problems, including
the switch from coal-to-oil bunkers, effectively withered away and lay
dormant until shortly after the Second World War. In 1950–1951 for example,
just 68,000 tonnes were shipped from Newcastle, New South Wales all of
which was destined for the "Pacific Islands" including New Zealand.

Throughout the 1950s various foundations were laid that were to have
spectacular benefits when they eventually paid off. The main developments
were: the post-war industrialisation of Japan (in particular its heavy steel
industry sector); the investment in exploration and resource development
in Australia; and the expansion of infrastructure requirements including
the ports.

During the 1960s and 1970s the metallurgical coal needs in Japan, and
the investment that flowed from Japan to Australia in terms of mine
development, built up a degree of mutual co-operation and beneficial
working practices. When the international demand for thermal coal first
impacted at the end of the 1970s, Australia was both perfectly placed and
poised ready to take advantage.

In 1983, Australia assumed the role of the world's largest seaborne coal
exporter. This leading position was maintained in 1984 and Australia has
remained the number one exporter of coal ever since.

There is no doubt that this solid (and rapid) progress has been built on
the back of demand by Japan; first by steel mills seeking high quality
coking coal, and latterly by the cement and electricity producers. However
this "vulnerability" to the whims of the Japanese buyer has of course been
appreciated for many years in Australia, but the absence of alternative
markets of such magnitude to where the coal could be transported
economically was drastically limited.

Much progress has been made with exports to the European market, but
because of the long haul involved it meant that either larger shipment sizes
are necessary (to obtain the economies of scale), or increased shipments
are made when freight markets were low. This occurred in the mid-1980s
when the delivered cost of the coal was competitive with shorter haul coals
from closer origins.

In recent years as other markets have opened up for coking coal (particularly in South Korea, and Taiwan), Australia has been well placed to take advantage of its proximity to these countries. It was a natural development therefore that when these same countries began to convert to coal from oil for electricity and cement generation, they should turn to their established coking coal supplier as the logical thermal coal provider. Nevertheless, despite this active move away from Japan, Australia still (1990) sends over 50% of its coal exports to Japan.

Table 42 Australia: coal exports (million tonnes)

Year	To:	Japan	%	Other	%	Total
1950/51		—	—	0.1	100%	0.1
1955/56		—	—	0.2	100%	0.2
1960/61		1.7	90%	0.2	100%	0.2
1965/66		7.6	95%	0.4	5%	8.0
1970/71		16.1	85%	2.9	15%	19.0
1975/76		23.8	78%	6.6	22%	30.4
1980/81		32.8	69%	14.6	31%	47.4
1984/85		43.9	52%	39.9	48%	83.8
1985/86		42.9	48%	47.0	52%	89.9
1989(Cal)		53.4	54%	45.3	46%	98.7

Exports

The spectacular growth in Australian exports since the early 1960s has been shown to be a combination of circumstance, opportunity and availability. However all of these elements could still have been present and yet the export industry need not have been half as successful if Australian coal had been too expensive to produce.

Whilst production costs vary from mine to mine and are not dependent upon coal quality, the bottom line is how cheaply coal can be mined, washed, transported, shipped and delivered to final destination. In this regard Table 43(a) shows the imported cif price of coking coal, as received, in Japan in yen per tonne. Table 43(b) shows the same imported price per tonne, but converted to a simple index such that all other origins are compared to the Australian price in each year, ie Australia 1960 yen 5,197 = 100.

It can be seen from these Tables 43(a) and (b) that Australian coking coal has always been less than the average price of all (including Australian) coals. Usually Australian coal has been about 20% less than the average price but on some occasions (1974–1975) it has been almost 40–50% cheaper than average. It is not appropriate to draw precise conclusions about the difference as many factors come into play; for example yen/dollar exchange rate, particular supply difficulties because of strikes, etc. Table 43(b) shows that in most years Australian coal is not much cheaper than Canadian coal and almost always dearer than South African coal.

Table 43(a) Japan: coking coal cif imports (yen/tonne)

Year	Australia	USSR	Canada	USA	S.Africa	All
1960	5,197	5,239	5,356	6,610	—	6,203
1965	4,832	5,213	5,118	6,662	—	5,664
1970	5,836	5,519	5,856	8,886	7,331	7,257
1975	11,789	15,390	14,812	22,293	12,971	16,600
1980	13,496	13,057	14,054	18,239	12,050	14,980
1985	13,032	12,871	16,117	16,422	11,842	14,287
1990	7,997	8,296	10,378	9,647	7,310	8,729

Source: The TEX Report—various years.

Table 43(b) Index: basis Australia cif price = 100

Yr	Australia	USSR	Canada	USA	S.Africa	All
1960	100	100.8	103.1	127.2	—	119.4
1965	100	107.9	105.9	137.9	—	117.2
1970	100	102.5	108.7	165.0	136.1	134.7
1975	100	130.6	125.6	189.1	110.0	140.8
1980	100	96.7	104.1	135.1	89.3	111.0
1985	100	98.8	123.7	126.0	90.9	109.6
1990	100	103.7	129.8	120.6	91.4	109.2

One factor of this relative but persistent cheapness, is as a result of the practice of Japanese purchasers investing in Australian mines. Long-term (10, 15 or 20-year) contracts with specified prices are then arranged. Whereas in the case of the other major Japanese supplier, the USA, purchases have been much less on a contract basis and more on a market related spot basis, and accordingly have often been higher priced because of the very nature of having to resort to that kind of purchase. There is also the virtually implied point that purchasers who are forced to resort to spot buying, are quite prepared to pay more to obtain the kind of coal they need.

Table 44 Australia: coal exports by state and type (million tonnes)

Year	NSW	Queensland	Coking	Steam	Total
1960	1.6	—	1.6	—	1.6
1965	5.6	1.5	6.9	0.2	7.1
1970	11.9	6.4	16.9	1.4	18.3
1975	14.5	15.4	26.5	3.4	29.9
1980	22.9	19.9	33.9	8.9	42.8
1985	40.7	47.2	49.8	38.1	87.9
1990	45.7	60.9	57.1	49.5	106.6

Source: Joint Coal Board/Queensland Coal Board.

During the 1960s and early 1970s, Australian coal exporters were very reliant on the Japanese market. In fact that reliance reached a peak in 1969 when 99% of Australian exports (some 16 million tonnes) was coking coal destined for Japan. Total coking coal imports by Japan have in fact been relatively stable since the mid-1970s as a consequence of the contractual arrangements and the evenness of Japanese crude steel production. The growth in steel production in newly industrialising countries (South Korea, Taiwan, etc.) has produced a convenient outlet for the surplus Australian coking coal that was expected to go to Japan.

Thermal coal exports by destination have, by and large, mirrored the trends of metallurgical coal. Most surprising perhaps is the importance of the European market, particularly in the mid-1980s, but this reflected the weakness in coal freight rates and the economies of that haul have to be carefully judged in the event of a major upturn in the freight market.

Table 45 Australia's coal exports by type and destination

Coking	1972	1976	1978	1980	1982	1984	1986	1988	1990
Japan	20.6	26.5	24.5	25.8	26.2	28.1	27.7	27.9	28.2
S.Korea	—	1.0	1.5	2.3	3.7	3.3	3.2	3.9	4.9
Taiwan	—	—	1.0	0.8	1.6	1.8	2.1	3.3	2.8
Oth Asia	—	—	—	0.4	1.6	1.2	4.1	6.3	6.4
Europe	2.0	3.7	6.2	4.0	4.0	9.4	11.8	11.8	10.6
S.America	—	—	0.2	0.1	0.1	0.9	1.0	2.0	1.6
Others	0.1	—	0.1	0.1	0.4	1.0	1.4	0.8	2.4
Coking	22.7	31.2	33.5	33.9	37.1	47.0	51.3	56.0	56.9

Thermal	1972	1976	1978	1980	1982	1984	1986	1988	1990
Japan	—	0.3	0.6	3.9	6.6	11.6	14.8	22.9	27.1
S.Korea	—	—	—	—	0.6	3.4	5.3	4.2	4.2
Taiwan	—	0.1	0.1	1.0	0.6	1.7	3.2	3.8	3.7
Oth Asia	—	—	—	—	0.6	3.1	4.6	5.8	4.6
Europe	0.9	2.6	3.6	3.9	3.9	8.5	12.2	6.6	8.7
S.America	—	—	—	—	—	—	0.1	0.2	0.2
Others	—	—	0.9	0.1	0.4	0.5	0.8	0.3	0.7
Thermal	0.9	3.0	5.2	8.9	12.7	28.8	41.0	43.7	49.2

Total	1972	1976	1978	1980	1982	1984	1986	1988	1990
Japan	20.6	26.8	25.1	30.1	32.8	41.0	42.5	50.7	55.2
S.Korea	—	1.0	1.5	2.3	4.3	6.7	8.5	8.1	9.0
Taiwan	—	0.1	1.1	1.8	2.2	3.5	5.3	7.1	6.6
Oth Asia	—	—	—	0.4	2.2	4.3	8.7	12.1	11.0
Europe	2.9	6.3	9.8	7.9	7.9	17.9	24.0	18.3	19.3
S.America	—	—	0.2	0.1	0.1	0.9	1.1	2.3	1.8
Others	0.1	—	1.0	0.2	0.8	1.5	2.2	1.0	3.2
Total	23.6	34.2	38.7	42.8	49.8	75.8	92.3	99.6	106.1

NB: Totals may not add exactly due to rounding.
Source: Joint Coal Board *Black Coal In Australia*, (various years) and *Australian Black Coal Statistics 1989*.

USA

Despite the great variation seen in the production level of bituminous coal in the US since 1914: from a low of 281 million tonnes in 1932 to a peak of 934 million tonnes in 1990, the exports-to-production ratio has remained broadly constant, never exceeding 15.5%, as in 1957, or falling below 2.4%, as in 1935. Notwithstanding this consistency, US coal exports have fallen into two distinct phases, and, more recently appear to have entered a third phase.

The two major phases correspond coincidentally to the inter-war and post-war eras. The third and most recent phase has occurred since 1979 with the growth of the international steam coal trade.

Phase 1 (1918–1939)

In 1914, bituminous coal production reached 383 million tonnes, of which 16 million tonnes was exported (11.7 million tonnes to Canada, 4.3 million tonnes to overseas destinations). This level of exports was fairly typical of the 1914–1918 period. Following the First World War, output in many European countries was much reduced due to manpower and machinery shortages and, as a result, demand for US coal leapt to a new record level of 20 million tonnes in 1920. In 1921, a coal miners' strike in the United Kingdom supported US exports temporarily, but for the period through to 1925 overseas exports returned to their "normal level" of about 4 million tonnes. In 1926 a miners' and general strike in the United Kingdom caused a severe short-fall in supply to overseas markets. The United States was able to demonstrate its ability to respond to a changing market situation and an exceptional level of exports were exported (19.6 million tonnes). The following year was artificially supported by contractual arrangements agreed to in 1926 and therefore exports were somewhat above a justifiable level. However the effect of the world-wide general depression and slump was already beginning to be felt. To put the extent of the slump into context: for the ten years 1930 to 1939 the aggregate of all the exports amounted to only 9.4 million tonnes—less than half the quantity exported in the single year of 1926.

The period therefore was characterised by a very consistent level of exports. If the two exceptional years of 1920 and 1926 are excluded then, for the 30 year period from 1914 to 1943, exports as a percentage of production varied only between 2.4% and 5.6%. The bulk of exports were destined for Canada which never took less than 7.6 million tonnes (1932) throughout the whole of the 1920–1940 period.

Reliable statistics of the use made of US coal in the country of destination were not compiled at that time, but an early generalised analysis was made[2] which showed that there was no particular emphasis on US coal for its metallurgical properties. Its use for domestic (household) consumption, the railroads, gasworks and general manufacturing were just as important.

Phase 2 (1946–1979)

While Phase 1 was characterised by a low but consistent level of overseas exports with Canada as the main export destination, Phase 2 typically had about double the percentage of shipments with the extra demand being accounted for by overseas (rather than Canadian) buyers.

The expansion in US exports correlated negatively with the decline in exports from the United Kingdom. However the demand for US coal at that time was based on the desire for good quality coking coals many of which were not available in the United Kingdom. This demand for coking coal was a natural consequence of the reconstruction programme started after the war and its inherent demand for steel. United Kingdom exports had suffered from the loss of their traditional overseas market, namely the supply of coal bunkers for cargo ships. Vessels switched increasingly to oil for propulsion (from almost complete coal use in 1913) to just half of their needs by 1925.[3] The USA, while taking over the role as the world's major supplier, was not competing in the same market because the market was changing and the United Kingdom could not supply the speciality coals required in this reconstruction period.

The coal exported to Canada has been mainly of thermal type destined for power stations use in Ontario to produce electricity, while the coal for overseas markets was principally of metallurgical type for the production of coke. Accordingly, the level of exports reflected international developments in steel production.

Phase 3 (post 1979)

Since 1979, the USA has been a leading supplier of thermal coal on the world markets. The surge in demand for thermal coal that started in that year, and was particularly apparent in 1980 with the loss of Polish supplies to the European market, was largely unexpected. Of all the major suppliers of coal in the world it was the USA alone that could respond to a changing market situation. This was because of the surplus capacity and the almost fragmentary nature of the US coal industry. A measure of the extent of this response can be seen by consideration of US thermal exports (excluding Canada).

2. League of Nations *Memorandum on Coal* Vol 1 (Geneva, May 1927).
3. J.P. Dickie *The Coal Problem* (League of Nations, Geneva, 1936).

In 1978 the United States exported just 0.3 million tonnes of steam coal. With the events in the Middle East, Poland and elsewhere, the US responded with 30 million tonnes of coal exports in 1981. By 1983, with the panic buying over, US exports fell back to 15 million tonnes. Since then the US suppliers have adapted to the needs of the global market and in 1990 some 28 million tonnes were shipped overseas.

It has been shown on at least four occasions this century: 1920, 1926, 1948–1955 and 1979–1981, that the USA alone has been able to act as a swing supplier to the world's coal consumers. Given the structural nature of the industry, which is primarily geared to large-scale steam coal production to cover domestic needs, there is no reason to believe that the country should not continue to fulfil this role for some time to come. This is because with such huge productive capacity there is nearly always some coal that is surplus to requirements in the system.

Table 46 USA: seaborne coal exports (excl Canada)[4]

Year	Thermal	%	Coking	%	Total	World	%
1965	7.7	25%	23.6	75%	31.3	59.0	53.1%
1970	4.8	16%	42.6	84%	47.4	101.2	46.8%
1973	1.6	12%	31.6	88%	33.2	103.8	32.0%
1975	5.0	11%	39.4	89%	44.4	127.4	34.9%
1979	2.3	6%	39.1	94%	41.4	159.4	26.0%
1980	14.5	22%	51.5	78%	66.1	188.4	35.1%
1985	19.9	29%	48.4	71%	68.3	270.9	25.2%
1990	28.3	35%	53.6	65%	81.9	341.0	24.0%

Throughout the 1970s the two principal areas of overseas coal exports for the United States were the Far East (especially Japan) and Western Europe (especially the EEC countries). Demand in both areas (up to 1979 at least) was determined largely by the requirements of the steel industry and limited by developments in their indigenous coal industries.

In addition, the USA, besides marketing to the large customers where shipments run into hundreds of thousands, or million of tonnes, also exports to many minor importers. This effort, which continues to the present day, often involves very small tonnage movements frequently of less than 100 tonnes. This policy could be said to be a special feature of the US industry with its proliferation of small producers.

The 1970s opened with an exceptional year with exports 12 million tonnes up on the adjacent years. The reasons for the greater than expected exports will be covered shortly, but the first characteristic that made it exceptional was the variation in the tonnage exported. The wide range of

4. 1965–1979: US Dept of Commerce, "US Bituminous Coal used for Metallurgical Purposes" Quoted in *International Coal* (various years) by NCA.

exports to be expected from year to year had, during the mid- to late-1960s largely come to an end and had settled at around 30–31 million tonnes annually. However throughout this period, despite the steadiness in the final export figure, the market was changing very significantly. In the five year period from 1966–1970, exports to Japan trebled from 7.1 million tonnes to 25.1 million tonnes as that country expanded its crude steel production to unprecedented levels. On the other hand in the EEC and other West European countries demand was falling dramatically. The effect on US exports facing increased competition from East European countries was that exports to Europe fell from 23 million tonnes in 1965 to less than 14 million tonnes in 1969.

Thus the main outlet for US coal shifted within a relatively short five-year period from Europe to Japan and distinctly towards metallurgical coal. The rapid expansion of the Japanese steel industry was not without its problems however, and its growth was one of the major contributing factors leading to the qualification of "exceptional" for 1970. The continued increasing demand for coking coal, combined with European mine closures, which in turn depleted stockpiles, led to a tightening of the supply situation in late 1969 and 1970. The situation did in fact lead to some isolated shortages, but the coal buyers, anticipating shortages, actually made the situation worse by ordering ahead of requirements in the spot market and succeeded in creating a "shortage scare".[5] The direct result was a surge in demand for US coal. In that year (1970) Japan increased its imports from the USA by 5.7 million tonnes whilst Europe, which for so long had been a declining market, increased its take by 5.8 million tonnes.

The following year (1971) revealed that there had indeed been an element of panic buying and exports declined by 12 million tonnes back to the level prevailing in 1969, although it should be recorded that Japanese steel production did fall from 93.3 million tonnes to 88.6 million tonnes. The over-buying of 1970 left traders with surplus stocks in 1971 and this in turn contributed to the decline in coal exports.

Although it was anticipated at this time that world coal trade would grow, there was a feeling within the US that security of supply among the consuming countries was of growing importance, and it was felt although exports would grow, market share would decline as Canada and Australia began their coal expansion programmes. In early 1972 the US Bureau of Mines assessed the long-term outlook as follows:[6]

"Although the United States was still the world's largest coal exporter in 1971 and exports are expected to grow, its share of the steadily expanding total world international trade in coal is expected to decline as consuming countries seek to diversify sources of supply and as much more competitive new coal mining

5. G. Markon "US Bituminous Coal Export Trends 1965–1974", US Dept of the Interior *International Coal Trade* Vol 44, No 3, March 1975.

6. "US Coal Exports in 1971", US Bureau of Mines, *ICT*, Vol 41, No 2, Feb 1972.

capacity expands. The larger world market and the search for increasingly diversified supplies at the lowest cost has led to larger consumer participation in coal mining, either in equity or in financing. The progressive expansion of large bulk carrier shipping units also point to a more stable, permanent international coal trade of the future. The larger fluctuations in demand and the buying of surplus coals for the immediate future are probably destined to account for a progressively smaller share of the market as more large-scale, long term coal contracts from assured suppliers come on stream during the balance of the decade."

It can be seen, with the benefit of hindsight, that this statement by the USBM was a perceptive assessment although spot purchasing continued throughout the two decades since then and has been stronger than that implied by the Bureau. Indeed the Japanese steelmakers (and others) have established the one year renewable contract as the "norm" in many agreements.

For 1972 and 1973, little change was seen in the export pattern of US coal; world coking coal supplies remained more than adequate, steel demand was still depressed in Europe, delivered coal prices were rising as a result of increases in freight rates and new suppliers were taking some traditional US markets; for example Australian coal was being shipped in to Italy.[7]

The immediate effect of the 1973 oil price rise on US coal exports was not upon the volume of coal traded but on price. In 1974, coal exports increased by only 3% compared to 1973 but the net value of exports more than doubled. The fob value jumped to $44.52 per tonne from $20.90 per tonne in 1973. Some of this increase can justifiably be accounted for by "pipeline" increases, which would have shown through anyway but undoubtedly of major significance was the heavy demand for uncommitted spot coals. There is also evidence that the export of lower quality coals that were not usually considered suitable for export, caused prices to rise to unprecedented levels.[8]

Exports to Japan showed a 42% increase over 1973 following the sharp increase in steel production seen in that year and also in 1974. However part of the increase in exports to Japan was due to anticipated shortfalls of supply from other sources, notably Australia where 2 million tonnes less was shipped. Coal exports to Europe were generally higher, but the uncertainty that prevailed immediately after the oil price jump of late 1973 was too close in terms of time to have any immediate effect on volume shipped.

In 1975, the established pattern of coal exports was in general much the same as for 1974. Shipments to Japan were down following a reduction in spot coking coal purchases while coal exported to Europe started to show

7. "US Coal Exports in 1972", *US Bureau of Mines, ICT*, Vol 42, No 2, Feb 1973.
8. "The International Coal and Coke trade of the US in 1974" *US Bureau of Mines ICT*, Vol 44, No.2, Feb 1975.

some signs of recovery as the region began to adjust to the shock of the oil price increase.

Since 1975, unquestionably the most significant development in US coal exports has been the shipment in major quantities of thermal coal to overseas destination. In the years from 1975 to 1978 US coal continued on its declining trend in the face of direct competition from prime export orientated countries such as Australia, Canada and, from 1976 onwards, South Africa. In addition, because the country was virtually a total coking coal exporter, the cutback in steel production in Europe and Japan reduced the demand for imported coking coal. A measure of this cutback can be gained by comparison of the two years 1973 and 1974 with the two subsequent years 1975 and 1976. For the first pair of years crude steel production in Japan and the EEC nine countries amounted to 542 million tonnes. For the second pair of years crude steel production in the same areas amounted to 456 million tonnes: a fall of some 86 million tonnes. The two subsequent years, 1977 and 1978 were at a similar depressed level.

The pattern of US exports over the period is shown below:

Table 47 USA: bituminous coal exports (excl Canada) (million tonnes)

Year	Japan	EEC-9	Oth Europe	Others	Total
1965	6.8	19.6	3.0	1.9	31.3
1970	25.1	15.1	4.4	2.8	47.4
1973	17.4	9.7	3.2	2.9	33.2
1975	23.1	13.3	3.9	4.1	44.4
1979	14.2	16.4	4.2	6.6	41.4
1980	20.9	29.3	7.3	8.6	66.1
1985	13.9	30.0	8.9	16.4	69.2
1990	12.1	----52.9-----		16.8	81.9

Source: US Bureau of Mines; US Dept. of Commerce; US National Coal Association; EIA.

In 1978, the US competitive position was further weakened by a 111-day strike by the United Mine Workers Association (UMWA). This strike was to have repercussions that were to reverberate for many years. Exports were reduced to such an extent that in March of that year US coal exports totalled less than 0.3 million tonnes compared to a monthly average in 1977 of over 4 million tonnes, in itself not a good year. The figure recorded in March 1978 was in fact the lowest of the decade. Subsequent to the strike a rapid recovery was seen but a further strike, this time by the clerks of the Norfolk and Western Railways, once again restricted exports from the Hampton Roads area, which is the most important export area for coal in the United States.

In 1979, the effects of the second oil price rise and the consuming countries first adjustments to the 1973 oil price rise increase, began to tell. The first major sector to convert to coal-fired usage where previously oil had been accepted was the cement industry, and in 1979 this conversion began to impact upon the demand for thermal coal. In fact thermal coal exports in that year came to 2.3 million tonnes of which 1.5 million tonnes went to Western Europe, principally West Germany, and 0.4 million tonnes was destined for Japan.

Nineteen-eighty was the first year of significant thermal coal exports from the United States. Throughout the year the effects of conversions of oil plants to coal firing increased coal demand but by August 1980 of greater significance was the disruption to the European customer of Polish supplies. Accordingly, exports from the US leapt to 14.5 million tonnes of which 12.4 million tonnes went to Europe. Once again in the space of a very short period of time a whole new market had opened up with an extra 12 million tonnes of thermal coal being required and the USA alone able to respond fully to the situation. Although it should be recorded that South Africa did increase its exports to the EEC by some 4 million tonnes, this was more of a capacity increase at the same time than a deliberate response to an expanding market.

The cycle of disruption to US coal caused by the renegotiation of employment contracts every three years and the memory of the devastating loss of supplies in 1978 made buyers apprehensive about a similar disruption for the next round of negotiations scheduled for Spring 1981. In fact a long strike did occur from 27 March to 8 June 1981, but this time coal exports were hardly affected and the result was that European buyers anticipating the loss of Polish and US supplies bought coal in record amounts to supplement stockpiles. In the event coal exports were barely affected and the worst month for exports, May, was still greater than any month in the previous strike year, 1978. Total seaborne exports reached a record 85.5 million tonnes of which 29.9 million tonnes was thermal coal.

It has been shown time and again that the ability of the US coal industry to respond to international situations is without parallel among the world's coal suppliers. That this is due to the highly competitive and fragmented nature of the industry is without question. In 1981 it was estimated[9] that there were nearly 6,000 mines in the United States. Further it was found that the UMWA's members account for only some 44% of domestic production with the remainder coming from non-union mines and mines affiliated with other unions. In addition, there are about 3,000 coal mining companies operating in 26 states; finally it was estimated that some 85 different companies produced in excess of 1 million tonnes of coal.

9. US Bureau of Mines Mine Safety and Health Administration. Published in "Facts About Coal— 1982". NCA, 1982 Washington.

As further emphasis of the US supply capacity it has been shown that coal exports account for only a small proportion of domestic production, typically about 10%. In addition, the country has a well established spot market system (this has been described in some detail in Gaskin's study[10]).

The new facilities that have been implemented in the last few years have added extra capacity to the terminals, but the problems of deep water ports remain. The congestion that was seen in 1981 has the potential to return due to the continuing inability to be able to load the largest vessels.

CANADA

Coal exports

Table 48 gives the breakdown of Canadian coal exports by area of destination since 1969. The trend away from Japanese dependency is clearly shown. From 1972 when 99% of Canadian exports were destined for Japan, through to 1990 when the proportion had fallen to 60%, there has been a continuous diversification of export regions.

Table 48 Canada: coal exports by destination (million tonnes)

Year	Japan	Europe	S.America	Other	Total
1969	1.1 92%	— —	— —	0.1 8%	1.2
1970	3.7 93%	0.1 2%	— —	0.2 5%	4.0
1973	10.6 97%	— —	0.1 1%	0.2 2%	10.9
1975	10.8 92%	0.8 7%	— —	0.1 1%	11.7
1979	10.6 76%	1.2 9%	0.5 4%	1.6 11%	13.9
1980	10.5 69%	1.1 7%	0.7 4%	3.0 20%	15.3
1985	18.5 68%	2.5 9%	1.0 4%	5.3 19%	27.3
1990	18.4 60%	2.9 9%	1.4 5%	7.8 26%	30.5

Source: Monthly Coal and Coke Statistics, Statistics Canada, Ottawa.

As there are substantial reserves of thermal coal in both Alberta and British Columbia there is considerable scope for increased exports of this coal grade. However, as is characteristic of free market conditions, it is one thing to have the coal available and another to have the customers for it.

It is readily apparent that the Japanese steel mills did not invest huge sums of money in Canadian mines simply because they wanted more coal—although there was certainly a strong element of this when the Japanese

10. M. Gaskin "Market Aspects of an Expansion of the International Steam Coal Trade" EAS Report No G2/81 (IEA London, England, August 1981).

Table 49 Canada: annual imports and net exports (million tonnes)

Year	Imports	Exports	Net Exports
1969	15.6	1.2	− 14.2
1970	17.1	4.0	− 13.1
1973	14.9	10.9	− 4.0
1975	15.2	11.7	− 3.5
1979	18.0	13.9	− 4.1
1980	16.1	15.3	− 0.8
1985	14.6	27.3	+ 12.7
1990	14.2	31.0	+ 16.8

Source: Monthly Coal and Coke Statistics, Statistics Canada, Ottawa.

Table 50 Canada: coal exports by type (million tonnes)

Year	Coking	Thermal	Total
1977	11.3	0.8	12.1
1979	12.8	1.1	13.9
1980	14.1	1.2	15.3
1985	22.4	4.9	27.3
1990	26.7	4.1	30.6

Source: 1977/79 Author's estimate.[11] 1980/90 Statistics Canada.

first went to Canada for extra supplies in the late 1960s. This view can be discounted simply because by consideration of the enormous problems associated with the Canadian coal industry. Not the least of these is the prevailing climatic conditions in the two western provinces. Mining conditions are generally acknowledged as hostile and necessitate export coals being dried before shipment. In addition the terrain is difficult with many major seams located over 1,000 metres above sea level.[12]

The transportation costs, which is a relatively fixed element, was estimated in 1984 at close to US$ 20 per tonne.[13] Accordingly, price reductions have to be achieved at the mine rather than at the port. This is obviously difficult given the hazardous conditions. It is also a fact that the Canadian economy is closely linked to that of the US economy and while Australia has devalued and the rand/yen rate has weakened substantially, the Canadian dollar/yen rate has weakened to a much lesser extent. In other words price differentials exist naturally in the delivered cost of coal to the Japanese market.

These price differentials have not, of course, stopped the Japanese steel

11. P. Rogers, "Developments in Seaborne Coal Trade 1960–85" (PhD Thesis, 1988).
12. M. Iwaki "Japanese Views on the Canadian Coal Industry", 7th Japan/Canada Coal Conference, Kobe, Japan, May 1984.
13. G.D. Coates "Thermal Coal—Issues Confronting Coal Exports", 7th Japan/Canada Coal Conference, Kobe, Japan, May 1984.

mills from buying Canadian coal. Therefore it is necessary to look beyond the price of the reasons for this. These appear to include the following:

1. Canada has been a reliable supplier of coal over the last two decades. Compared to innumerable strikes in Australia, Canada has been relatively strike free.

2. Japanese interests have taken substantial equity interests in the mines themselves, therefore they are buying their "own" coal.

3. Because of the location of the coal mines it is unlikely that future exports will be constrained by domestic demand conditions.

4. Canada has developed very modern deep water ports such as Roberts Bank and Prince Rupert ideally suited to the largest vessels to minimise the sea transportation cost.

5. The very strong desire of the Japanese to diversify their supply sources.

Conclusion

The legacy of the Japanese steel industry's massive investment over the years is an extremely modern port development at Roberts Bank and Prince Rupert; an extensive infrastructure that has opened up the interior of the western provinces; positive benefits to the economy with coal contributing about US$ 1 billion a year from exports and, finally, an 'in place" infrastructure exploitable both for other commodities and, more importantly, to enable coal exports to other countries to take place, in particular South Korea and Taiwan.

POLAND

Exports

Up to 1990 coal exports from Poland were controlled by "Weglokoks", a "foreign trade enterprise" company, which claims the title of the largest exporting organisation in the world for coal, coke and brown coal. Weglokoks was founded as an independent enterprise on 2 January 1952, and concerns itself with all aspects of the marketing and export of Polish coal including transport, finance and technical support. The organisation maintains overseas offices in several of the main consuming countries including Eire, Finland, Denmark, United Kingdom, Italy, France and Brazil. Weglokoks has representative offices in all of the country's coal exporting ports and has a specialist chartering company—"Polfracht Shipbroking and Chartering Co."—which attends to the shipbroking requirements of the organisation.

Polish coal first appeared on the world market in the early 1920s, although sources differ on the actual year: Weglokoks claim 1922,[14] while Polfracht claim,[15] 1924 for its "first appearance". In any event Polish coal established itself very quickly on the world scene such that by 1926 annual exports had leapt to 15 million tonnes, and was being shipped to such diverse destinations as the USA and Australia, as well as the more obvious markets of France, Austria and Scandinavia.

The coal being exported at that time was used mostly for steam railways, ships bunkers and domestic needs. Table 51 shows the principal uses to which exported Polish coal has been put in recent years in non-Socialist countries.

Table 51 Consumption of Polish coal exports by sector (million tonnes)

Year	Power	Steel	Industry	Domestic	Total
1979	14.2	7.7	3.1	1.4	26.4
1980	10.9	6.0	1.9	1.3	20.1
1981	3.0	3.3	1.0	0.6	7.9
1982	6.7	5.4	1.1	1.0	14.2
1983	7.7	6.3	2.3	1.3	17.6
1984	12.4	6.8	3.6	2.0	24.8
1985	9.5	7.2	2.4	1.8	20.9
1986	7.1	7.0	1.7	1.4	17.2

Source:[16]

Polish coal types

Traditionally Poland has been a major exporter of thermal coal and it is only relatively recently that coal mines and reserves were exploited that could supply the world metallurgical coal market. In 1965, these exports amounted to 640,000 tonnes with none of this coal type exported between 1945 and 1960.

The reason for this development was a natural desire to expand markets further. This was particularly in the face of the growing demand for metallurgical coal that was evident on the world market at the end of the 1960s, and also as an outlet for coking coal beyond the requirements of the Polish steel industry, which by 1970 was producing 11 million tonnes of crude steel.

14. Polish Coal Review *Special Number on the Occasion of the Traditional Barburka Day* Weglokoks, Katowice, Poland 1982.

15. A. Sygalt "Poland Coal Production and Export Capabilities", Conference Paper presented at "Coal Trans 81".

16. Weglokoks; "Polish Coal Review". Various years. Katowice, Poland.

Polish coal markets

It can be seen that in regional terms, Western Europe is by far the most important market with only about 6% of exports moving outside the European area. Of these non-European destinations, Brazil is the most important buyer taking 1.7 million tonnes of coking coal in 1983.

This trade distribution pattern is not wholly unexpected in terms of the central geographical location of Poland both to East and West Europe. Shipments to more distant locations are not unknown however, as already mentioned; and recently the principal long haul customer, other than Brazil for which a mutually beneficial bilateral trade agreement exists for Brazilian iron ore in exchange for Polish coking coal, was Japan.

In the period 1969 to 1979, cumulative imports by Japan of Polish coking coal amounted to 10.5 million tonnes, although shipments were on a declining trend from 1974 onwards, and finally ceased in 1981 with a single Panamax shipment of 65,000 tonnes.

Table 52 Exports of Polish hard coal (thousand tonnes)

Year	Socialist Steam	Coking	Capitalist Steam	Coking	Total Steam	Coking	Total
1945	5,379	—	162	—	5,541	—	5,541
1950	15,332	—	10,867	—	26,543	—	26,543
1955	16,170	—	7,757	—	24,027	—	24,027
1960	8,493	—	8,983	—	17,496	—	17,496
1965	11,915	84	8,478	562	20,390	646	21,036
1970	12,143	—	11,445	5,225	23,588	5,225	28,813
1975	14,115	521	13,528	10,184	27,643	10,705	38,348
1979	14,490	475	18,483	7,915	32,973	8,390	41,363
1980	10,300	528	14,371	5,750	24,671	6,278	30,949
1981	6,517	594	4,768	3,137	11,285	3,731	15,016
1984	14,549·	3,685	17,609	7,233	32,159	10,918	43,077
1985	12,028	3,243	13,398	7,516	25,426	10,760	36,186
1990	10,106	585	7,415	9,908	17,521	10,493	28,014

Source: Statistics from CHZ "Weglokoks", Katowice, Poland.

Table 53 Polish hard coal exports by area (million tonnes)

Year	Socialist	Europe	Other	Total
1970	12.1	15.3	1.4	28.8
1975	14.6	21.2	2.5	38.3
1979	15.0	24.0	2.4	41.4
1980	10.8	18.5	1.6	30.9
1981	7.1	6.9	1.0	15.0
1982	14.3	12.6	1.6	28.5
1984	18.2	22.5	2.4	43.1
1985	15.3	18.7	2.2	36.2
1990	10.7	14.6	2.7	28.0

Source: Statistics from CHZ "Weglokoks" Katowice, Poland.

Future outlook for Polish coal production and exports

With the introduction of martial law in 1982 the coal industry stabilised its production and made a very rapid recovery to close to 1979 levels by 1984. However, with an international debt estimated[17] at $44 billion as at the end of 1983 (Bank of England estimate), prospects for further expansion (whether rapid or not) were highly restricted. This is because although there is adequate capacity in terms of surplus coal available for export, the debt restrictions meant that this capacity could not be developed economically without a further massive injection of capital which the country did not, and does not, have available.

Towards the end of 1984, which was the peak year for exports, it was reported[18] that there were six hard coal mines under construction, three existing mines being expanded, 14 active mines under further development and sufficient investment to "keep production of 49 other active mines at their present level".

Today (1991) the problems of the Polish coal industry can seem almost overwhelming. As we have said the industry is of vital importance to the country because of the hard currency it generates, but at the same time the industry faces many difficulties mainly as a result of an inability "to increase the economically reasonable extraction".[19] The International Energy Agency cites[20] the following factors that will lead to a decline in Polish coal mining:

"The geological conditions systematically worsen . . . the increase of mines depth, the increase of rocks temperature, the growing share of thin and variable thickness layers, the localisation of coal in safety pillars, large gas contents of new deposits, inflow of salted waters. The easy accessible resources exhaust. The numerous mines will be fully exhausted in near future."

Accordingly, there seems little prospect within the remainder of this century for a coal production level much in excess of 145 million tonnes. This in turn implies, given the domestic needs of the country, an export level of no more than 30 million tonnes annually, and the distinct possibility that exports may drop to 20 million tonnes a year.

SOUTH AFRICA

The international political problems that faced the government of South Africa over the last 30 years meant international isolation from some

17. Economic Progress Report, "International Debt", United Kingdom Treasury No169, July 1984, London.
18. A. Smiglewski "Poland as a Producer and Exporter of Coal", paper presented at "CoalTrans 84", London, October 1984.
19. IEA *Energy Policies Poland 1990 Survey* (Paris 1991).
20. *Ibid.*

important coal trading partners (eg Denmark, the Netherlands) over some of that time. External oil supplies were extremely difficult to obtain and, with no natural indigenous supplies available, the country was forced to develop its own alternative forms of energy security; in general these have centred on coal. Oil is produced from coal under some of the most advanced technological processes at the SASOL plant. Otherwise coal is used as a prime fuel input for the railways, power stations and boilers for steam raising. In addition, South Africa is able to exploit its need to obtain coal for domestic purposes, and at the same time maintain some of its trading links— thereby obtaining valuable foreign currency— by allocating some of this coal for export.

In the mid-1980s coal production for the domestic market was controlled by three large mining houses (Anglo American Coal Corporation— Amcoal, General Mining Union Corporation— Gencor, and Rand Mines).[21] When the Richards Bay coal terminal was opened in 1976 exports were handled by three domestic groups (the Transvaal Coal Owners Association (TCOA); the Natal Associated Collieries (NAC); and the Anthracite Producers Association (APA)). Since 1979 this tripartite group has expanded to include several multinational oil and mining companies.

Official statistics of coal exports from South Africa by country of destination have not been issued for many years. Nevertheless total coal exports on a monthly and annual basis are compiled by the Minerals Bureau of the Department of Minerals and Energy Affairs. In addition, the Chamber of Mines in their annual statistical tables summarise the production and domestic disposal (including exports) of South African coal production.

In fact coal has been exported from South Africa for most of this century, but it was only comparatively recently that the country began to exploit the several advantages that it has. The real spur to this expansion was the 1973 oil price rise which, within a few months, created the conditions for a new potential demand for thermal coal. The South African authorities decided to exploit this expectation (of increased demand), and within three years the port of Richards Bay was opened. This port, which has direct access to the coal fields via the dedicated rail lines, has transformed the position of the country in the rankings of the world's exporters such that South African exports rose from 2 million tonnes in 1975 to almost 50 million tonnes in 1990.

Some coking coal is exported from South Africa and has found its way in the past into both the European (especially French) and Japanese markets. Since about 1981 nearly all of the coking coal has been destined for Japan originally under a 14.5 year contract for Witbank Coal that started in late 1972. This contract has specified 2–2.5 million tonnes of coking coal

21. R.L. Cohen "South Africa Coal Exports: an Overview", paper presented at "CoalTrans 84", London, October 1984.

annually since fiscal 1979, although actual shipments, as recorded by the Japan Ministry of Finance, have reached 4.6 million tonnes (1984).

In the first instance, annual coal exports from South Africa are decided by the government on an allocation basis, which specifies the quantity permitted to be exported in a particular year by an individual producer. This measure was introduced by the South African authorities as a means of ensuring that domestic producers were not squeezed out of the export market when oil/energy companies started to take an active interest in the country's mineral potential. If an export permit is not granted by the authorities then participation in the export market is not possible.

The rapid growth in exports also benefited from a few natural advantages that some other countries have either not been able to exploit fully, or were not open to them, including:

1. The rapid time for mine construction estimated at only three to four years compared to nearly ten years in the UK, for example.

2. The concentration of coal in relatively easy working conditions enabling large-scale, open-cast sites to be fully mechanised with the ensuing benefits of scale, etc.

3. A relatively cheap labour force compared to say, Europe, Australia or the USA.

4. The geographical location of South Africa makes it uniquely placed to obtain access to both European and Far Eastern markets.

5. Available deep-water facilities at a modern port designed from the "bottom up" for the new largest sizes of bulk carriers.

6. The ability to expand port, rail and mine facilities.

7. The political will to actively encourage coal exports.

The above points represent some of the major reasons for the internal success of motivating the coal industry towards exports. From the consumer's point of view the cheapness of South African coal (coupled with its ready availability and the opportunity to benefit from the economies of scale attributing because of the deep water facilities) considerably helped the commercial decision. In addition, the skilled marketing expertise of reputable international oil companies, meant that customers' needs were established by a network of offices within Europe rather than trying to market the coal from a South African base.

Against these positive factors, has to be set the political anguish that some countries/companies have felt in importing coal from that country. By and large these negative factors have not constrained the available supply and only since the mid-1980s did they have any significant effect.

In addition, the vulnerability of South African exports to disruption through sabotage, etc, has always been a real problem since the country decided to concentrate on one major port. On the other hand the authorities are well aware of this strategic weakness and have taken steps to protect all aspects of the export operation.

The official abolition of the apartheid laws in June 1991 and the USA's lifting of sanctions in July 1991, should open up new potential for South African coal producers. It remains to be seen if purchasing companies will face any political restrictions. Certainly the opportunity is there, and many coal importers are known to be keen to take advantage of the cheaper prices that South Africa has to offer.

Table 54 Distribution of coal trade in South Africa (million tonnes)

Year	Export	Power	Industry	Other	Total
1963	0.93	19.78	6.65	14.35	41.71
1965	0.77	22.75	7.92	16.37	47.81
1970	1.29	29.45	8.75	14.96	54.45
1975	2.29	41.33	12.13	13.26	69.01
1976	6.31	43.51	11.87	13.04	75.09
1980	28.44	58.85	16.13	14.15	117.57
1981	29.88	58.29	22.95	16.73	127.85
1985	44.30	62.40	44.10	15.20	166.00
1989	46.79	68.70	45.45	14.56	175.50

Sources: 1963–1980: Fuel Research Institute—Pretoria. 1981–1989: Minerals Bureau, Dept of Mineral and Energy Affairs.[22]

NB: Owing to differences in classification, the Minerals Bureau figures are not directly comparable with those formerly issued by the Fuel Research Institute.

UNITED KINGDOM

Tables 55 to 57 record the supply, and consumption of the UK coal industry for the period. Detailed analysis of the UK coal industry has been covered elsewhere by other authors, notably: Buxton,[23] Asteris[24] and Cox[25].

Despite the numerous "Plans for Coal" and revisions that have been issued since the National Coal Board was nationalised in 1947, it remains the case that most have been wide of the mark. This does not necessarily reflect any deficiency on the part of the forecasters—rather the very complicated and ever changing nature of the energy market.

In recent years, particularly since 1960, the UK has played only a relatively minor rôle in terms of coal availability in the European area. However, as far as the European coal balance is concerned, its rôle has

22. Both sources quoted in *Statistical Tables* (The Chamber of Mines of South Africa, Johannesburg, 1987 & 1990).

23. Neil K. Buxton, *The Economic Development of the British Coal Industry* (Batsford Academic, Edinburgh, 1978).

24. M. Asteris "Britain and the European Coal Trade 1913–39" (University Library ✓ Birmingham. M.Soc.Sc. Dissertation, 1971).

25. A. Cox "Future Strategies for Coal in the United Kingdom", University of Newcastle-Upon-Tyne, England (PhD thesis 1986).

been much more important. The figures in Table 55 explain the dramatic change that was seen at the start of the 1980s:

Table 55 United Kingdom: coal

Year	Exports	Imports	Net Exports
1981	9,113	4,290	+ 4,823
1985	2,432	12,635	− 10,203
Change	− 6,681	+ 8,345	− 15,026

The cause of this huge swing in the coal balance in the United Kingdom was caused by the year-long National Union of Mineworkers' strike, which started in March 1984. It can be seen that although imports in the period increased by only 8.3 million tonnes, the net difference is some 15 million tonnes. As most of the UK exports are destined for the European area, this area had to try and supplement its traditional UK supplies either from indigenous production, which was not practical, or from increased imports.

Table 56 United Kingdom: coal supply (million tonnes)

Year	Production	Of which: Deep	Open Cast	Net Imports	Imports	Exports
1972	121.8	109.1	10.4	+ 3.2	5.0	1.7
1973	132.0	120.0	10.1	− 1.0	1.7	2.7
1975	128.7	117.4	10.4	+ 2.9	5.1	2.2
1979	122.4	107.8	12.9	+ 2.2	4.4	2.2
1980	130.1	112.4	15.8	+ 3.5	7.3	3.8
1984	51.2	35.2	14.3	+ 6.6	8.9	2.3
1985	94.0	75.2	15.6	+ 10.2	12.6	2.4
1990	93.2	71.5	17.8	12.3	14.7	2.5

Source: UK. Dept. of Energy: "Energy Trends"— Monthly Bulletin.

Table 57 United Kingdom: coal consumption by sector (million tonnes)

Year	Total	Power	Coke	Industry	Domestic	Other
1970	155.6	75.8	29.9	19.1	20.6	10.1
1973	133.4	76.8	21.9	12.1	14.5	8.1
1975	122.2	74.6	19.1	9.7	11.6	7.3
1979	129.4	88.8	15.1	9.2	10.5	5.8
1980	123.5	89.6	11.6	7.8	8.9	5.5
1983	111.5	81.6	10.5	7.2	7.9	4.4
1984	77.3	53.4	8.3	6.0	6.4	3.3
1985	105.4	73.9	11.1	7.5	8.6	4.2
1990	108.4	82.5	10.8	7.4	4.9	2.8

Source: UK Dept. of Energy: "Energy Trends"—Monthly Bulletin.

A recent analysis[26] found that substantial changes had occurred in the industry in recent years, such as a doubling of productivity over the 1987–1991 period; and a massive reduction in the number of pits to 65 employing some 57,000 mineworkers as at March 1991. At the same time there had been substantial investment amounting to over £7 billion from 1979 to 1991.

The report recognises that once coal capacity is lost it is virtually irretrievable and says: "Reductions in the size of the industry are easy to make, whereas expansion at a later date would be extremely expensive and difficult." It remains to be seen what future is chosen for the UK coal industry.

OTHER EXPORTERS

Soviet Union

Until recent times Soviet coal production has been aimed primarily towards the internal needs of the country; secondly, towards the needs of other CPE countries; and thirdly, to the non-Socialist export market. With the proximity to customers in the West European region accessible by rail, some exports, for example to Finland and Austria, have arrived by that means. Equally, most of the shipments to the CPE area have also gone by rail and are not included in the totals for seaborne coal trade, the simplest definition of which includes shipments to Japan, the EEC, other Western Europe (excluding Finland and Austria), and "Others".

As far as the West European market is concerned, the USSR now ranks as a minor coal supplier. The European Community, for example, imported just 4.4 million tonnes from the USSR in 1990. The country has some potential as a supplier to the Far Eastern market. Japan in particular has shown interest in the Kuznetzky coking coal deposits. In addition, South Yakutsk coal, although located over 2,000 km from the export port of Nakhodka, is part of a long-term trade agreement between the two countries.

Table 58 USSR: coal exports by destination (million tonnes)

Year	EC	Oth Europe	E.Europe	Japan	Other	Total
1973	3.7	2.7	11.5	2.8	3.8	24.5
1975	3.8	3.2	11.7	3.2	4.2	26.1
1980	2.8	3.6	15.8	2.1	1.2	25.5
1985	1.3	5.4	12.8	3.9	1.6	25.0
1990	4.4	3.6	15.4	8.7	6.7	38.8

Source: UN-ECE and author's estimates.

26. UK House of Commons, Energy Committee, Fifth Report: 'Clean Coal Technology and the Coal Market after 1993'', 10 July 1991.

People's Republic of China

Exports of coal from China reached a new record in 1990 when over 17 million tonnes were shipped overseas. Hard coal has been exported since before 1980, but it is really only since that time that the volumes have been making a significant impact on world trade. Currently China is the world's largest coal producer and consumer, but as can already be implied, some 98% is consumed within the country.

A recent United Nations study[27] estimated that coal production might increase to 1,200 million tonnes annually by the year 2000 and that exports at that time could exceed 20 million tonnes. The coal is exported mainly to Japan and other countries in the South East Asian region. Some coal does come on to the European market and in 1990 this was estimated at 2.5 million tonnes. It has been estimated[28] that China could be exporting anywhere between 25 and 50 million tonnes by 2000, but much will depend on the pace of infrastructure development.

Colombia

In 1976—initiated by the oil price rise of 1973—a company "Carbocol" was created in Colombia under the auspices of the Ministry of Mines and Energy. The purpose was to create, develop and evolve a coal industry for the country. The main project centred around the El Cerrejon coal deposit located in the Guajira department.

While the coal deposits have been known about for many years, the coal was never exploited for use in Colombia because of the availability of hydro-electric supplies located closer to the concentrations of population. As a result the El Cerrejon coal deposit was developed solely with the intention of being used on the developing international steam coal market.

Currently (1991) the pit is producing some 15 million tonnes annually with plans to expand to 25 million tonnes by the end of the century—if demand merits it. The coal is low sulphur and low ash and suited to the growing environmental concerns that have rightly arisen in recent years over the future of burning fossil fuels.

The coal is transported by rail car some 140 km to the port of Puerto Bolivar where it can be loaded into capesize vessels up to 170,000 deadweight. Because of its strategic location it is well placed to serve the markets of Europe and the Southern United States. Shipments have been made to the Asian market, but it is the north Atlantic consumers that are seen as the main customers for the future.

The importance of coal to the Colombian economy should not be under-estimated: since 1985 the industry has accounted for about 15% of GNP,

27. P. Rogers *World Coal Trade up to the Year 2000* (UN-ECE, Geneva 1990).
28. G. Doyle *China's potential in international coal trade* (IEA, London, 1987).

and in 1990 was responsible for 10% of trade exports by value. The coal industry in Colombia employs some 20,000 persons most of whom are involved in the El Cerrejon project.

Venezuela

While in 1991 Venezuela remains a small exporter of coal the potential does exist for the country to move into the ranks of the more established suppliers during the 1990s. Most of the interest centres on the Gusare basin, which is estimated to contain some 350 million tonnes of "proven" reserves with a further 8,000 million tonnes of "indicated" and "hypothetical" reserves. While the Gusare project is the main focus there are several other deposits—Santo Domingo, Naricual, Lobatera—that also have sizeable reserves.

Currently Venezuela is exporting some 1.5 to 2 million tonnes per year through the Port of Maracaibo. Petroleos de Venezuela SA, the state-run energy company is hoping to raise exports to the 6–7 million tonne level during the 1990s. However the infrastructure requires a large investment estimated at between $400 and $500 million.[29] Because of this large capital requirement joint venture companies are being formed to overcome the problems especially the bottle-neck at the draft restricted port.

Indonesia

Exports from Indonesia have been promised throughout most of the 1980s, but now in the 1990s those expectations, which so often in the past have failed to materialise, do at last seem set to take off in a very rapid manner.

Table 59 Indonesia: production, consumption and exports

Year	Production	Cons:	Electricity	Cement	Exports
1985/86	2.07		0.21	0.47	1.02
1986/87	2.75		0.47	0.62	0.97
1987/88	3.48		1.75	0.85	0.99
1988/89	5.18		2.04	0.94	1.54
1989/90	9.48		4.60	1.48	2.69
1990/91	11.20		4.80	1.50	4.70

Source: Directorate of Coal.

Official plans[30] in Indonesia call for the tripling of coal production by the mid-1990s. Production is divided amongst government companies, state

29. *New York Journal of Commerce* "Venezuelans Expanding Basin Mining Activity", 15 July 1991.

30. PT Tambang Batubara Bukit Asam, quoted in *Financial Times* "International Coal Report", 28 June 1991.

controlled contractors and private companies. Most of the growth is expected to be generated in the contractor group.

Amongst the proliferation of over 30 companies (including the state run organisations) there are three that are of special significance:

1. PT Kaltim Prima: This is the largest project with production scheduled to reach 7 million tonnes by 1995.

2. PT Arutmin: The second largest project, which could hit 6 million tonnes by 1993.

3. PT Adaro Indonesia: Five million tonnes of coal production scheduled for 1995, but the main interest is for the projects "Envirocoal". This coal claims very low ash content (1%) and negligible sulphur (0.08%).

CHAPTER 5

The importers

JAPAN

Raw materials supply—domestic

The requirements for pig iron, coke and fuel oil in the steel-making process explain the fluctuating demand for metallurgical coal. In addition to the over-all demand by the steel industry, imports have been significantly affected by the continuing background of a declining domestic mining industry.

Domestic metallurgical coal production reached a peak in the five-year period 1940–1944 when total output aggregated some 275 million tonnes—an average of 55 million tonnes annually. After the war the devastation of Japanese industry led to a rapid decline such that by 1946 just 20 million tonnes were produced. Working under difficult, thin seam conditions, the Japanese coal mining industry was gradually re-built to the extent that the average output during the 1950s was 45 million tonnes annually.

The post-war peak occurred in the 1960–1964 period when an average 53 million tonnes was produced annually. The increasing availability of cheaper supplies from abroad, the exhaustion of thin seams, and the generally difficult mining conditions has meant that the industry has been in a state of initially rapid, but more recently slow, decline.

The last time 50 million tonnes of coal was produced in Japan in a single year was 1966, but by 1970 output had slumped to less than 40 million tonnes and then to less than 20 million tonnes by 1975. By the early 1980s a decision was taken to try and stabilise coal production at 18–20 million tonnes for the foreseeable future.[1] However even by 1984 output slipped

Table 60 Japan: domestic coal production (A) (thousand tonnes)

Year	Coking	Thermal	Total
1970	12,290	27,404	39,694
1973	11,460	10,954	22,414
1975	9,484	9,515	18,999
1979	7,793	9,850	17,643
1980	6,943	11,084	18,027
1985	3,921	12,464	16,385
1990	111	8,151	8,262

1. Agency for National Resources and Energy *Long Term Energy Supply and Demand Outlook in Japan.* (16/11/83).

further to less than 17 million tonnes. In the end the struggle to maintain even a residual metallurgical coal mining capacity was lost and production finally ceased in March 1990.

Table 61 Japan: imported coal (B) (thousand tonnes)

Year	Coking	Thermal	Total
1970	48,767	1,405	50,172
1973	55,823	1,031	56,854
1975	60,706	1,401	62,107
1979	56,118	2,436	58,554
1980	61,816	6,412	68,228
1985	70,143	22,845	92,988
1990	74,051	31,370	105,421

Table 62 Japan: total coal supply (A + B) (thousand tonnes)[2]

Year	Coking	Thermal	Total
1970	61,057	28,809	89,866
1973	67,283	11,985	79,268
1975	70,190	10,916	81,106
1979	63,911	12,286	76,197
1980	68,759	17,496	86,255
1985	74,064	35,309	109,373
1990	74,162	39,521	113,683

Thermal coal

Thermal coal imports did not become a feature of the Japanese market until 1974 when the decline in supplies from the domestic mines forced Japanese consumers to look abroad for their coal. Until that time anthracite had been the principal non-coking coal import. Thermal coal had been used quite extensively in Japan up to the early 1970s both for electricity generation and in the domestic sector, but the principal source for these groups was indigenous coal supplies.

Table 63 Japan: thermal coal consumption (thousand tonnes)[3]

Year	Power	Chemical	Cement	Paper	Others	Total
1980	9,599	133	5,384	173	2,207	17,496
1985	24,543	1,152	8,149	755	1,337	35,936
1989	28,335	1,898	7,710	1,765	2,747	42,455

2. Tex *Coal Manual* (Tokyo, Japan, various years).
3. IEA *Energy Statistics* (OECD, Paris, 1991).

The closure of extensive parts of the coal-fired electricity generating capacity in the early 1970s led to massive declines in the consumption of coal. In 1970, 19 million tonnes of coal was consumed by power stations, just three years later annual consumption was down to 7.5 million tonnes and stayed at that level until 1979 when oil-to-coal conversions began to impact on total requirements.

Electricity sector

The conversion to coal firing (from oil) and the construction of new coal-fired plant—triggered by the oil price rise of 1973—first began to materialise in 1979. That year not only represented the low of the last 30 years in terms of thermal coal consumption, but also marked the revival of coal as a viable source of electricity production. The further oil price rise that was to occur later in the year was seen as merely confirming that the right decisions had been made.

The growth in electricity demand that had been characteristic of the 1960s and early 1970s, had been anticipated by the electricity producers. They put into operation a programme of nuclear power generating plant that started to come to completion about the time of the second oil price increase. In the two-year period 1979–80 for example, new nuclear generating capacity coming on stream totalled 2,834 MW with an additional 2,460 MW of hydro-electric plant. Most of the effort directed towards coal-fired plant at this time was concerned with the starting of their construction. The government controlled EPDC (Electric Power Development Company) commenced the construction of (at the time) the country's largest coal-fired utility at Matsushima (2 × 500 MW), which came on stream in fiscal 1980 and 1981. By the end of fiscal 1982 some 40 coal-fired units of 5,762 MW were in operation, and 13 coal-fired units with a total, capacity of 4,193 MW were either under construction or being converted from oil.

Table 64 Japan: fuel consumption for electricity production[4] (MTOE)

Year	Coal	Oil	Gas	Nuclear	Hydro	Total
1970	16.52	44.39	1.21	1.13	19.63	83.88
1973	12.00	68.52	2.04	2.38	17.63	102.57
1975	12.41	63.84	4.52	6.16	21.07	108.00
1979	11.58	70.57	13.04	17.25	21.05	133.49
1980	10.54	60.63	15.88	20.23	22.78	130.06
1985	20.89	40.69	26.80	39.10	21.92	149.40
1989	23.29	46.99	29.48	47.39	7.05	88.25

Includes auto-producers of electricity.

4. OECD "Energy Balances of OECD Countries", (Paris, 1987 & 1991).

Table 65 Japan: electricity generated by fuel type (1,000 GWhr[5])

Year	Coal	Oil	Gas	Nuclear	Hydro	Total
1970	60.1	210.2	4.5	4.6	80.1	359.5
1973	37.3	340.8	10.5	9.7	71.9	470.2
1975	42.3	301.9	20.3	25.1	86.0	475.6
1979	44.5	315.9	73.0	70.4	85.9	589.6
1980	50.0	269.6	81.1	82.6	93.0	576.3
1985	99.9	195.1	128.2	159.6	89.5	672.2
1989	116.2	253.3	148.1	182.9	89.2	791.2

Table 66 Japan: electricity generated by fuel (%)

Year	Coal	Oil	Gas	Nuclear	Hydro	Total
1970	16.7%	58.5%	1.3%	1.3%	22.3%	100%
1973	7.9%	72.5%	2.2%	2.1%	15.3%	100%
1975	8.9%	63.5%	4.3%	5.3%	18.1%	100%
1979	7.5%	53.6%	12.4%	11.9%	11.9%	100%
1980	8.7%	46.8%	14.1%	14.3%	16.1%	100%
1985	14.0%	27.2%	17.9%	26.2%	14.7%	100%
1989	14.7%	32.0%	18.7%	23.1%	11.3%	100%

Cement industry

Despite the major effort being put into the conversion from oil to coal-fired electricity plants in the early 1980s, the main industrial sector that was first able to switch fuels was in fact the cement industry. When oil had a price advantage over coal there was little point in using coal other than for specialised needs. Accordingly coal, which had been the principal fuel used in cement manufacture up to the early 1960s, suffered a decline after that time.

Coal and oil are used as energy sources in the kilns and to switch from one fuel to another is a relatively easy matter of changing the burning equipment. Naturally there are other associated requirements such as tonnage sites and waste disposal, but with coal the by-product is coal ash, which does not create significant problems as ash can be integrated into the clinker. Thus within a relatively short space of time the change over was completed.

It was the surge in import demand for thermal coal (from 1 million tonnes in 1979 to 10 million tonnes in 1981) largely for the cement producers that triggered global demand for thermal coal trade. Imports for power stations in significant volumes only exceeded 2 million tonnes for the first time in 1981. This surge was caused by an extra requirement of 4.7 million tonnes

5. *Ibid.*

Table 67 Japan: fuel consumption by cement industry[6] (MTOE)

Year	Coal	Heavy Oil
1970	—	7.89
1973	—	10.07
1975	0.47	7.82
1979	1.10	9.18
1980	3.02	6.30
1985	4.86	1.56
1989	4.71	3.91

by the cement industry and 2.4 million tonnes by the conversion of existing oil-fired power stations to coal burning.

Supply of coal to Japan

It has been mentioned previously that Japan was aware of its dependence upon "outside" suppliers for its energy sources. The rational answer, as far as the Japanese coal importers were concerned, was to take an interest (sometimes financial, sometimes technical) in the coal mines in the supply countries. Subsequently, the policy has been to arrange contracts ranging from five to 20 years at agreed tonnage levels and prices to ensure security of supply. Naturally, for various reasons, for example strikes, accidents, changes in world demand etc, the tonnages required from year to year have sometimes exceeded and sometimes has been less than the "correct" tonnage. Japanese buyers were well aware of this factor and have used their position on the world market to cover the difference.

For example 75% of a buyer's requirements may be covered on an existing contract with 25% bought on a spot shipment basis. When demand is more than expected the buyer turns first to the contract supplier to take more tonnage: in many cases the contract supplier can supply some additional volume but not all of it. In such cases the Japanese buyer turns to the spot market—particularly in the USA, which has a large quantity of spare capacity and numerous suppliers competing to provide tonnage. When demand conditions are weak, spot purchases are stopped and the contract supplier will be asked not to provide the full quota. In many cases of weak demand the contract supplier is not in a position to argue with the principal customer and either accepts the contract volume cut or tries to re-negotiate a price compensation clause. In the circumstances this is difficult and not always successful.

6. *Ibid.*

Supply sources

Over the last couple of decades there have been several clear trends in the imports of coal by Japan:

1. The emergence of South Africa from 1973 onwards when it first regularly supplied coking coal.

2. The emergence of the People's Republic of China as a thermal and coking coal supplier to Japan from 1975 onwards.

3. The expansion of Australian exports— originally as a coking coal supplier but since 1980 as a thermal coal supplier. In 1970, Australia supplied one-third of Japanese imported coking coal needs, by 1990 this proportion had grown to 46%.

4. The erratic pattern of supply from the USA. Although the normal pattern of supply was disrupted by a severe strike in 1978, it is clear that despite that exceptional year the USA exports to Japan have been the most reliable of all major suppliers. The reason for this variation has in turn been led by identifiable factors:

(a) The use by Japan of the USA as a spot supplier of coking coal.

(b) Because of the vulnerability of the US supplies to exogenous circumstances like the strength of the US dollar against other currencies.

(c) The different coal qualities within the US has meant that in many cases when a particular quantity was required on a spot basis, the US was the only available supplier.

POWER STATION COAL TRADE

This section attempts to quantify the gestation of seaborne coal trade destined for consumption in thermo-electric power stations. It concentrates on developments over the evolution of the trade from 1973 to 1983; it will be seen that most of the expansion has occurred since 1979.

Data sources

Thermal coal imports specifically for power stations use are rarely identified separately as a category in the international statistical publications dealing with coal trade although some detailed work has been done in this respect.[7] Fortunately it is possible to come reasonably close to the figures involved by analysing production and consumption patterns in each country and by careful interpretation of the available import data.

---

7. P. Rogers "International Seaborne Trade in Power Station Coal", Association des Ingenieurs Electriciens Sortis de L'Institut Montefiore, Liege, Belgium, October 1985.

Some organisations, such as the European Commission, publish annual tables showing coal imports by origin as supplied to public power stations. However, as will be shown later, the interpretation of this "source" data has not always proved accurate. For example, it is obviously necessary to remain within the confines of total coal imports, which generally is a well established and accepted figure. The next problem is to separate the quantity of coking coal imports. The International Energy Agency in its annual report[8] differentiates between coking and steam coal. Other sources are available. There is, for example the United Nations, which produces an annual energy review,[9] and the OECD, which has done work on consolidating energy statistics both within[10] and outside[11] the OECD member countries group.

Many difficulties can occur, for example, as mentioned previously, the Statistical Office of the European Commission records, via a questionnaire to each member country, the thermal coal imports for use in public power stations. When the government energy department of a certain country was contacted by the author in the course of querying an import figure that was believed to be suspect, it was established that the principal importer had not been contacted to find out the quantities. The calculation (of power station coal imports) had been made by simply deducting the coking coal figures from the total import figures and assuming that all of the balance was for power station consumption, thereby totally ignoring any coal for the household and industrial sector.

There are other pitfalls, which are cited here so that the interested reader can avoid them. It is well known that Rotterdam is used as a transshipment port. In 1984 the United Kingdom had to import coal from a variety of sources because of the miners' strike, and the Netherlands is recorded in the United Kingdom trade statistics as supplying 2.2 million tonnes, whereas of course there is no coal production in Holland.

When comprehensive aggregate data is not available, the next step is to approach the country authorities involved. Again this often causes difficulties as for political reasons governments were often reluctant to reveal the destination of their exports—for example South Africa and the People's Republic of China; or their import sources—for example South Korea, which does not identify coal from South Africa.

In yet other countries although the exports or imports are recorded the coal type is not defined precisely, or even at all. For example, often there is no differentiation between coke, anthracite or bituminous coal in some official trade statistics. Or, if bituminous coal is recorded, coking and thermal types may not be sub-divided. Of course even in the ideal situation when all

8. International Energy Agency *Coal Information Report* (OECD, Paris).
9. UN *Yearbook of World Energy Statistics* (New York, USA).
10. IEA *Energy Statistics—1971/81* (OECD, Paris, 1983).
11. IEA *Energy Balances of Developing Countries—1971/82* (OECD, Paris, 1984).

types are precisely and accurately recorded this is still only part of the picture. There is nothing to stop coking coal being burnt in electric power stations as virtually any hard coal will do, and equally, even if a country is a large thermal coal importer it does not necessarily mean that any of it is used in the production of electricity. This is particularly true in Japan where thermal coal for power stations use exceeded other uses only as recently as 1983. Until that time the cement industry had been the largest importer.

The methodology that is used here is to assume that all *imported* coking coal is destined for use by the steel makers in the production of coke. All the remainder is assumed to be thermal coal, perhaps including anthracite as appropriate; this is then analysed by each country to derive a power station use figure. This analysis was done by consideration of the coal consumption figures for each country, which in some cases proved difficult. For example, the Republic of China's (Taiwan) figures do not appear in any UN publications despite the "world" total often designated therein. In other cases the analysis is relatively easy. For example some countries did not have any coal-fired electricity capacity—ie Singapore, Malaysia, or Hong Kong—before 1981, but in general only very detailed and careful analysis revealed the extent of coal use. In several cases, particularly those countries that had their own indigenous production, it is necessary to contact the import organisation, or the power station company, to obtain reliable data.

Finally, in the very rare cases where no response can be obtained in this way, a judgement can be formed based on total imports and the consumption of coking coal for steel making. In addition, statistics on the *exports* of coal to each country were consulted as even if coking exports were used in power stations, it would not be possible to use ordinary thermal grades to produce coke. Also it should be remembered that in most cases this absence of definitive data affects only a very small number of countries and the tonnages involved were frequently of 0.2 million tonnes or less; relatively large in absolute terms, but in the context of global seaborne coal trade very marginal in the effect on the distribution between coal types. It must be accepted therefore that inevitably there is a degree of error within the total trade figures which is difficult to quantify precisely but, based on a subjective assessment of the methods used, it is hoped that the gross error might be less than 2% of total trade in any year.

Global developments

The long explanation of data sources described above is necessary to establish fully the credibility of the data. In the broadest context it will be seen that historically the European Community (at least up to the end of the 1970s) was virtually the only major market for imported power station coal, and indeed thermal coal. While the balance of consumption is still held by Western Europe as a whole, the fastest growth since 1979 has been in the countries of the Far East, particularly Japan, Taiwan, Hong Kong and South Korea.

It can be seen from Table 68 that total seaborne power station coal trade in 1973 was less than 10 million tonnes annually. By 1983, this total had increased almost sixfold to nearly 60 million tonnes and further to over 80 million tonnes in 1985. For comparison purposes, total seaborne coal trade in 1973 was around 103–104 million tonnes while total world coal trade, ie including land movements was 175 million tonnes.[12] By 1983, total seaborne coal trade had risen to 200 million tonnes[13] with total trade estimated at 260 million tonnes.

From the time a decision is made to construct a new power station to the date of its coming on stream is, on average, eight to nine years which naturally varies from country to country (according to an estimate made by UNIPEDE[14]). It is not surprising therefore that since the 1973–1974 oil price rise it was only in the 1980s that a number of utilities were commissioned.

It might be thought therefore (from inspection of Table 68) that the expansion in power station coal *trade* simply reflects the decline in indigenous production in certain countries rather than any new growth. While this may be a contributory factor in a few countries, this thought can be largely dispelled by consideration of the volume of total world production of coal. This is currently of the order of 3.6 billion tonnes.[15] As only around 10% of total output enters world trade, this implies of course that 90% is produced and consumed in the country of origin. Even if the "big three" producers: the USA, the USSR and the People's Republic of China are excluded, current production in the rest of the world still amounts to some 1.1 billion tonnes. In 1973, total world coal production amounted to 2.2 billion tonnes and if the same three countries are excluded, output in the rest of the world came to 790 million tonnes, an absolute increase in these other producers of 310 million tonnes from 1973 to 1990, clearly implying a rapidly expanding coal industry world wide.

Much of this additional output has come from producers such as Australia, Canada and South Africa, who are highly geared towards the export market. It can be argued that early impetus to the thermal coal trade as a whole was given by the cement producers who could relatively easily switch to coal as a power source, but most of that switch was complete by 1982–1983 and further expansion in that area is clearly limited by prospects within the construction industry.

12. United Nations, *Ibid.*
13. Author's estimates.
14. UNIPEDE "Programmes and Prospects for the Electricity Sector—1983/88" Sept 1983, Paris.
15. Author's estimates published in ICR *Coal Year* (1991).

Table 68 Seaborne coal trade developments thermal coal (million tonnes)

Year	Power Station Coal Trade	Other Thermal Coal	Total Thermal Coal
1973	9.8	9.2	19.0
1974	14.0	13.4	27.4
1975	20.5	10.0	30.5
1976	20.9	10.2	31.1
1977	24.5	14.1	38.6
1978	27.9	12.3	40.2
1979	35.7	18.3	54.0
1980	51.0	22.9	73.9
1981	49.0	37.1	86.1
1982	53.9	35.0	88.9
1983	55.1	32.5	87.6
1984	67.3	40.5	107.8
1985	81.3	48.9	130.2
1986	86.0	48.9	134.9

Source: Author's estimates.

Table 69 Seaborne coal trade developments: coking coal (million tonnes)

Year	Seaborne Coking Coal	Total[16] Seaborne Thermal + Coking	Total Trade[17] (Inc Land Movements)
1973	84.8	103.8	175.2
1974	91.7	119.1	191.5
1975	96.9	127.4	193.5
1976	95.7	126.8	191.5
1977	95.4	134.0	202.0
1978	85.9	126.1	202.7
1979	107.3	161.3	228.8
1980	113.9	187.8	258.1
1981	122.4	208.5	271.1
1982	119.9	208.8	268.7
1983	112.0	199.6	266.0
1984	130.8	238.6	304.7
1985	140.7	270.9	335.8
1986	135.0	269.9	336.1

Source: Coking coal: author's estimates.

16. Fearnley *World Bulk Trades* (Oslo, 1973–1976).
17. 1973–1978: UN *Yearbook of Energy Statistics* 1979–1986: V.Calarco *The Coal Situation* (Chase Manhatten Bank, Annual).

Regional developments

European Community

In the EEC countries, power station coal imports from outside the Community grew at an annual rate of 9.2% for the period 1975–1983. Thermal coal imports for non-power station use increased substantially in percentage terms, but in absolute tonnage measure did not increase very much (+ 2.6 million tonnes from 1975 to 1983) reflecting the decline both in domestic and industrial consumption which as a sector fell from 69 million tonnes annually in 1971 to 34 million tonnes in 1981. In the face of such a decline it is perhaps more remarkable that any increase in imports was seen at all, but this reflected the closure of indigenous capacity rather than a wholesale switch to coal.

Third party imports specifically for consumption in public power stations were limited in the initial period after the 1973–1974 oil price rise by three factors:

1. the capacity of coal-fired generating plant;
2. import policies of member countries (such as the requirement by the UK's Central Electricity Generating Board to take its coal from the National Coal Board); and
3. the inability of producer countries to deliver at cheaper than indigenous price.

Single-fired generating capacity, ie coal only, has been on a slowly declining trend since 1973 when it stood at 58,373 MW. By the end of 1983 this total had fallen to 53,315 MW, which is not at all a surprising development as the priority of electricity producers has been security of supply of fuel and the ability to meet peak load demands.

Dual-fired capacity on the other hand declined sharply after 1976 when much of the Italian bivalent capacity was re-rated as oil only. However from a trough of 79 GW at the beginning of 1978 coal-fired dual capacity stations increased to 88 GW by the beginning of 1984. Of course it is one thing to have the installed capacity and yet another to use it. Analysis of the utilisation of the coal-fired stations by the simple yardstick of comparing net output with net capacity confirms quite clearly the extra use that has been made of these stations. From a theoretical low of 35% in 1974 and 1975, utilisation since 1980 has averaged around 55% of the maximum possible in a year, although these figures should be interpreted with caution due to the different timings within an individual year that such capacity becomes available.

Table 70 shows imports of power station coal together with imports of "Other Thermal" coal. This "Other" section would include imports of coal by the cement industry. Unfortunately, available statistics for the European Community do not split coal consumption of the cement plants between imported and indigenous coal.

Table 70 EEC: thermal coal imports (million tonnes)

Year	Power Station	%	Other Thermal	%	Total
1975	17.4	76.3%	5.4	23.7%	22.8
1976	17.8	74.5%	6.1	25.5%	23.9
1977	21.6	77.4%	6.3	22.6%	27.9
1978	23.6	82.8%	4.9	17.2%	28.5
1979	29.9	79.9%	7.5	20.1%	37.4
1980	40.7	82.6%	8.6	17.4%	49.3
1981	38.0	78.7%	10.3	21.3%	48.3
1982	38.3	77.8%	10.9	22.2%	49.2
1983	34.5	79.9%	8.7	20.1%	43.2
1984	39.6	72.7%	14.9	27.3%	54.5

Table 71 EEC: power station coal supplies by source (million tonnes)

Year	Indigenous Supply	Other EEC	3rd-Party Imports	Total Supplies
1975	98.1	2.0	17.4	117.5
1976	109.1	1.8	17.8	128.7
1977	110.7	4.0	21.6	136.3
1978	115.0	5.7	23.6	144.3
1979	119.1	5.0	29.9	154.0
1980	122.1	5.6	40.7	168.4
1981	122.6	7.0	38.0	167.6
1982	130.7	6.0	38.3	175.0
1983	128.0	4.5	34.5	167.1
1984	74.6	2.5	39.6	116.7

Source: Eurostat

It can be seen from Table 71 that indigenous supplies of coal have continued to make a valuable contribution to power station coal requirements. While third party imports doubled from 1975 to 1983, in no single country was domestic coal supplying less to the power stations in 1983 than it was in 1975. Therefore it is more true to say that the cut-back in productive capacity seen in the Community in the period meant that there was not the spare capacity available to meet the extra demand caused by the completion of new generating plant.

Other Europe

These countries include the group under the definition of OECD Europe, less the EEC member countries. In this category coal development has been generally slow and mixed. Total hard coal consumption in thermo-electric power stations doubled in the 11 years from 1971 to 1981. In 1971 total consumption for the group was 6.3 million tonnes with 60% of this being consumed in Spain. By 1981 consumption was 12.6 million tonnes but of this total Spain's contribution has risen to a massive 84%. Finland and Turkey are the two other main consumers in the group but after reaching a peak consumption of 4.7 million tonnes in 1980, the switch by Finland to a base load electricity generation produced by nuclear plants led to the most dramatic change in the consumption patterns of any country in Western Europe.

In that year Finnish consumption of hard coal is recorded as 4.0 million tonnes, by 1981 this had fallen to less than one million tonnes, and by 1983 barely a quarter of a million tonnes of hard coal was being consumed in the production of electricity. The change was caused partially by unusually good hydro-electric output in that year, but principally by the rapid commissioning to full power of the four nuclear power plants, such that nuclear and hydro, which in 1978 accounted for 37% of electricity generated, had by 1983, risen to 75% output.

In Spain, which for the uniformity of the statistics presented here is included in "Other Europe" rather than the "European Community" data, changes in energy policy resulting from revisions to the national energy plan have in the past led to some very erratic forecasts of coal requirements. In addition indigenous production of coal and lignite, and estimates of the different contributions of each solid fuel type have added to the difficulties in forecasting the precise requirements of imported coal in any particular year. Sharp reductions in the outlook for Spanish nuclear generation have re-drawn the solid fuel demand balance.

Table 72 Other OECD Europe coal imports: 1973–1985 (million tonnes)

Year	Coking	Power	Other	Total
1973	7.5	1.8	2.1	11.4
1974	8.2	2.0	3.1	13.3
1975	9.2	2.9	1.7	13.8
1976	9.2	2.9	1.5	13.6
1977	8.0	2.7	3.0	13.7
1978	7.3	4.1	1.5	12.9
1979	9.8	4.6	0.9	15.3
1980	9.4	6.0	1.8	17.2
1981	9.0	4.0	6.9	19.9
1982	9.4	2.5	8.3	20.2
1983	9.3	2.1	7.9	19.3
1984	10.0	2.5	8.7	21.2
1985	8.1	5.7	3.7	17.5

Far East

In 1973, imports of hard coal for consumption in power stations were negligible, certainly less than one million tonnes and with a high probability of being less than half a million tonnes. In some countries coal was being imported for this use but often in quantities of a few hundred or thousand tonnes. One notable exception was Taiwan, which had variable requirements depending on the output of its domestic supplies. For example, in 1970 the country imported 100,000 tonnes but none at all for the following two years. In 1973 it imported 36,000 tonnes and then 190,000 tonnes in 1974. In the four years after that, ie 1975–1978, it imported just 20,000 tonnes. However, by 1980 the total had leapt to 2.8 million tonnes with the start up of new coal fired capacity particularly the Hsinta 2 × 500 MW stations which commenced operations in 1982 and 1983.

Elsewhere the only other significant importer in the 1970s was Japan, but even there imports were less than might have been thought as they did not commence until 1974 and from then until 1979 did not exceed a quarter of a million tonnes in any year. In 1980, however, the first batch of conversions from oil-to-coal began to take effect as did the completion of new units. With a declining indigenous coal production over the last decade there was only one source of replacement tonnage available and that of course was imports, which in 1980 increased to 1.5 million tonnes and by 1984 had reached 8.6 million tonnes. By the end of 1984 Japan had effectively completed its conversion from oil programme and virtually all new capacity was to be provided by new construction.

In South Korea imports for power station use did not commence until shortly before completion of the Samch'onp'o 560 MW No 1 unit in 1982. In that year requirements for imported coal reached 1.1 million tonnes, which subsequently increased to an estimated 4.3 million tonnes by the end of 1984.

In Hong Kong, coal imports for consumption within the country are only destined for use in the new power stations, which makes allocation of imports relatively straightforward. The completion of the Castle Peak "A" station in March 1982 meant that imports that year were 1.5 million tonnes—there was in fact some pre-stocking in 1981—but this amounted to just 0.1 million tonnes. Subsequently, with the completion of new units, imports have increased rapidly such that in 1990 imports reached 10 million tonnes.

Other importers
The only other major importer of power station coal is Israel which began importing in 1981 in support of its new Hadera installations when 0.5 million tonnes were received. Since then imports have grown rapidly to 4.1 million tonnes in 1990 with further new units planned for the 1990s.

Table 73 Far East: power station coal imports (million tonnes)

Year	Japan	S.Korea	Taiwan	H.Kong	Total
1973	—	—	—	—	—
1974	0.2	—	0.2	—	0.4
1975	0.2	—	—	—	0.2
1976	0.2	—	—	—	0.2
1977	0.2	—	—	—	0.2
1978	0.2	—	—	—	0.2
1979	0.2	—	1.0	—	1.2
1980	1.5	—	2.8	—	4.3
1981	3.6	—	2.8	0.1	6.5
1982	5.1	1.1	2.6	1.5	10.3
1983	7.5	3.3	2.0	3.4	16.2
1984	8.6	4.4	4.9	4.5	22.4
1985	11.8	9.0	7.0	5.5	33.3

Conclusion

Tables 68 and 69 demonstrate quite clearly the rapid expansion that there has been in power station coal trade particularly since 1979. Future growth prospects will depend on the same factors that determined the development of the trade to date, namely: power station construction programmes, government energy import policies, the price of oil, the energy mix in a particular country, and last but not least developments in the indigenous coal industry of a consuming country. The supply of coal from exporting nations is not seen as a limiting factor for the foreseeable future.

Plate 1 Lee-Norse continuous miner at an underground coal face.

Plate 2 A rotor excavator at the Luchegorsk coal cut pit.

Plate 3 Mid-stream loading in the Mississippi.

Plate 4 Dalrymple Bay Coal Terminal stockpiles and HPS stockpiles.

89

Plate 5 Maasvlakte terminal at Rotterdam.

Plate 6 A rotor excavator with a capacity of 1,000 tons of coal an hour at work in a coal-field in Ekibastuz.

PART II

Carriage of Coal

CHAPTER 6

The ships employed to carry coal

HISTORY OF THE BULK CARRIER

The seaborne coal trade has a long history; the first record of coal being carried to London was in the fourteenth century. Before the railways were built the only way to transport coal was by sea, as road transport depended on the packhorse and the small loads that these animals were able to carry. At that time the roads were little more than tracks so that travelling was a slow and unreliable process, the result of which was that transport was restricted to the more costly items whose price included what must have been a substantial transport cost.

When the industrial revolution caused an unprecedented increase in the demand for coal, the best method of transport was by sea. For many small seaside towns coal was delivered by sea before even their harbours were built, the cargo was discharged by the simple expedient of beaching the ship at high tide and when the tide had fallen sufficiently the cargo was discharged into carts and taken the short distance to its destination in the town. By the time the next high tide refloated the ship, the cargo was discharged and she was ready to resume her voyage. The apparent simplicity of this procedure concealed its dangers, many ships were destroyed when a sudden gale sprang up before they were able to get clear of the beach.

Up until the middle of this century London was heated by "sea coal", to distinguish it from charcoal. This was coal mined in the coalfields of Northumberland and Durham and transported to London by sea. The voyage along the North Sea coast of England, before the days of adequate charts or navigation aids, with foul weather and constantly shifting sandbanks was particularly hazardous. It is not surprising therefore, that many of the ships employed in this trade came to an untimely end.

These harsh conditions came to be an excellent proving ground for both ships and men. Captain Cook was born in Whitby and went to sea at an early age, and the experience which he gained navigating in these difficult waters was to prove invaluable during his later expeditions. The ship chosen by Captain Cook for his discovery of Australia, the *Endeavour*, was originally an East Coast collier, a type of ship with which he would have had many years of experience.

The early steam engines operated at low pressures and consequently suffered from a high coal consumption. This was because the reciprocating engines of this period had a low efficiency and consequently a high coal consumption. Thus there was a requirement for coaling depots on all the major shipping routes so that ships could replenish their bunkers at frequent intervals as their high coal consumption prevented them from carrying enough coal to last the voyage. This in turn led to a demand for ships to service these depots. During the latter half of the last century and the early years of the present century there grew up a large fleet of "tramp" steamers whose existence depended to a large extent on outward cargoes from the coal ports in South Wales. Having loaded this first cargo, they were then in a strong position to compete on the open market for further cargoes, eventually returning to the UK with a cargo of bulk raw materials for the growing economy or a cargo of grain to feed the working population.

Plate 7 The ST *Portwey* is one of the last surviving coal-burning tugs. She was built in Glasgow in 1927 for the Channel Coal Company to tow coal barges between Portland and Weymouth.

The traditional tramp steamer lasted until the 1950s but it could not compete with the larger bulk carriers that were coming into service about this time. Their ability to carry more cargo at lower cost enabled bulk carrier owners to offer shippers freight rates that were substantially below the costs of the traditional tramp ship. Tramp ships or "tweendeckers", as they are known as today, are still around in considerable numbers but they are no longer used in the major bulk trades.

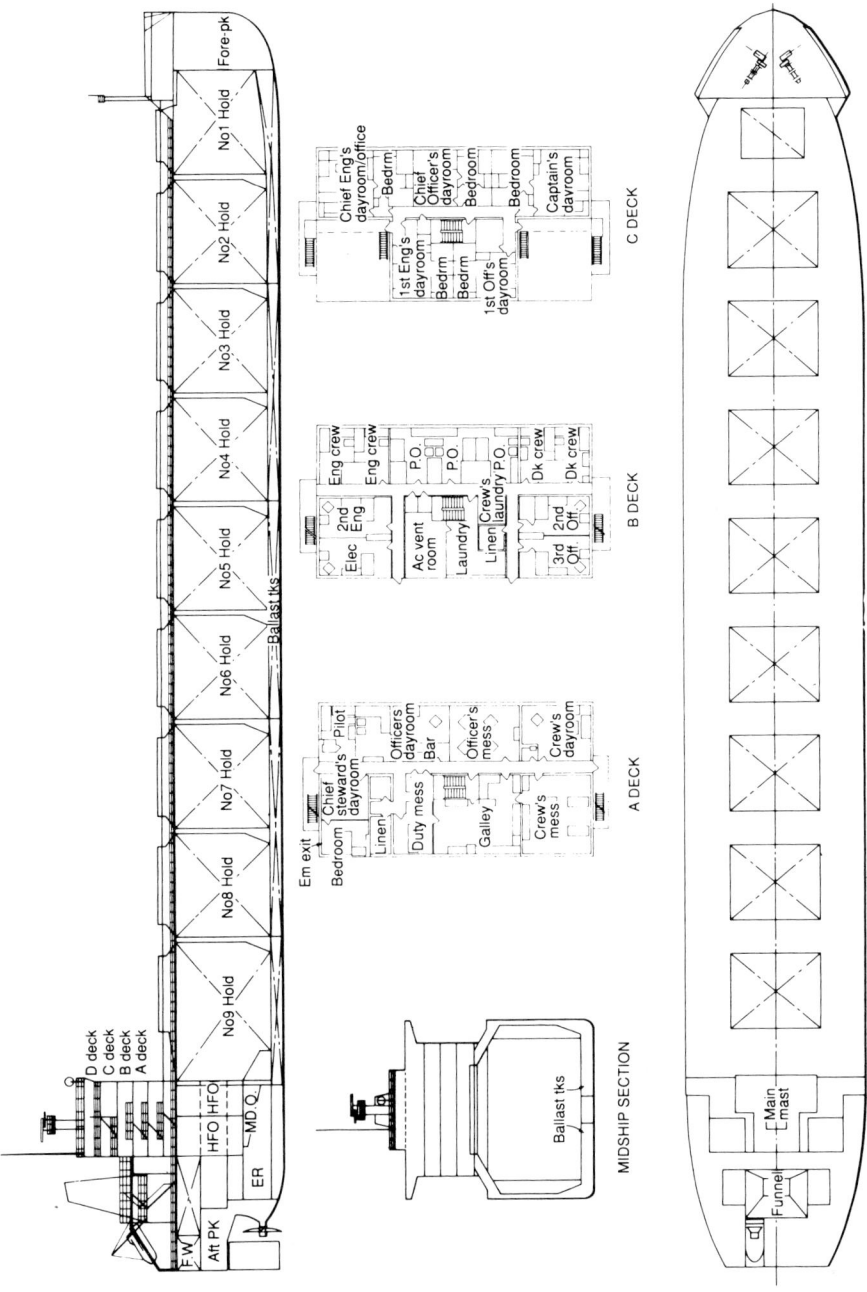

Figure 1 A modern Panamax bulk carrier—the *Solidarnos*—built by Burmeister & Wain for the Polish Steamship Company.

Although frequently considered a post-Second World War design, the bulk carrier owes its origins to designs dating from the last century. The use of water ballast, which can be discharged without interrupting the loading of cargo (an important feature of all bulk carriers), was first found in a ship built in 1854. A self-discharging bulk carrier with the ability to discharge coal at 800 tons per hour appeared in 1912.

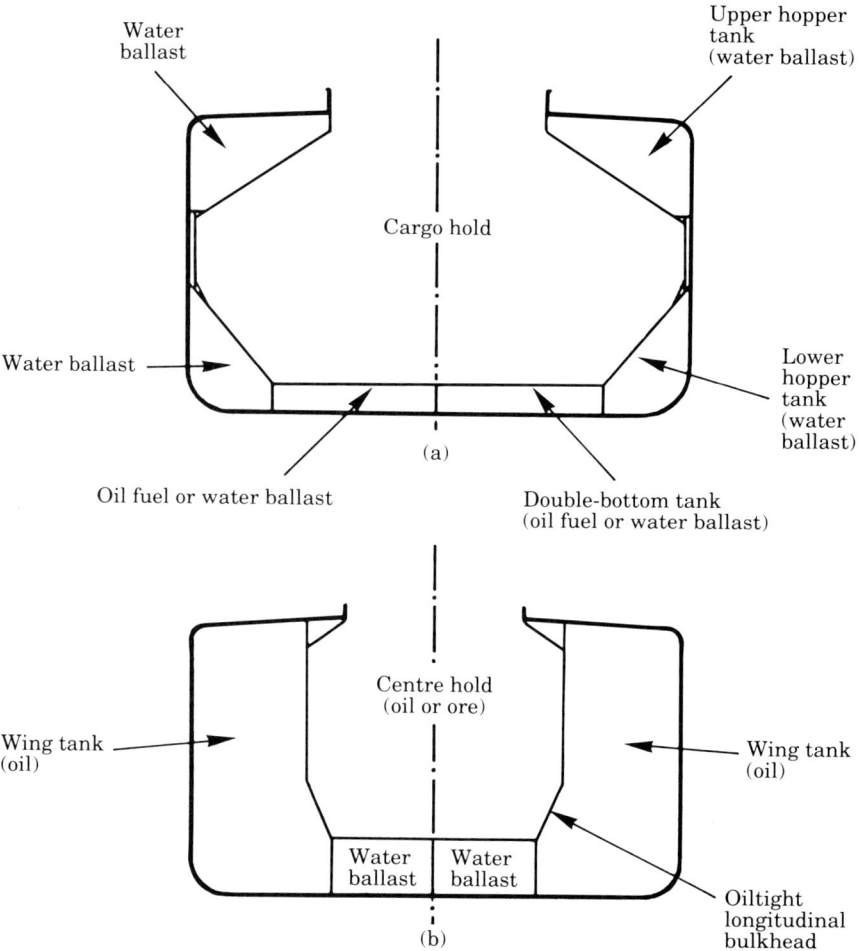

Figure 2 Transverse sections of (a) a bulk carrier and (b) an ore/oil carrier showing the tank structure.

The bulk carrier was developed to provide economies of scale in the transport of dry bulk commodities. It differed from the tramp ship in having only one deck, thus providing a large clear hold suitable for grab discharge. The engine was moved from the centre of the ship to the stern to provide a clear centre section over for the carriage of cargo. The disadvantage of this layout is the excessive trim by the stern when the ship is empty, which is compensated for by large water ballast tanks, the largest of which, the forepeak tank, is at the bow. The ship will have between five and about nine holds, one or more of which may be floodable to provide space for water ballast. The main feature of this is that they will be clear of any obstruction to facilitate rapid cargo handling. The actual shape of the hold will vary depending on the type of ship. The ship that is intended for bulks and packaged timber and possibly containers will have a hold with square sides to facilitate the stowage of these cargoes.

The ship that is intended for bulks only will have holds designed to make the cargo self-trimming. This is important for grain and some types of coal where movement in a seaway could endanger the ship. To achieve this, water ballast tanks are installed in the top and bottom corners of the hold. The top tanks have the effect of reducing the width at the top of the hold rather like a funnel. This will reduce the space available for the cargo to shift so eliminating the danger of cargo movement and at the same time ensuring that as the cargo settles during the voyage, any void spaces in the main part of the hold are filled by the cargo in the funnel.

When discussing the bulk carrier it is important to distinguish between the ore carrier and the general purpose bulk carrier. The bulk cargoes carried today have a wide range of stowage factors ranging from 15cu ft per ton for iron ore to 50 cu ft per ton for grain, with coal stowing at about 40cu ft per ton. This means that a ship that is intended to be a general purpose bulk carrier will need to have large holds to cover a wide range of possible cargoes while an ore carrier will have small holds that must be specially strengthened for the high density cargo. Generally today, the ore carrier is a large ship, usually Capesize or 150,000 + dwt while the smaller size ships are the general purpose bulk carriers. The statistics on bulk carriers include a number of ships that are unsuitable for cargoes of coal because their holds are too small and would be full of coal before loaded to their maximum deadweight.

The type of bulk carrier in use today can be grouped into the following classes:

(a) *150,000 +*. These are the largest bulk carriers in use today. Their trade is mainly with Japan carrying iron ore and coking coal for the steel industry.

(b) *Capesize*. These range from 80,000 tonnes deadweight (dwt), ie just too large for the Panama Canal to the maximum size of around 150,000 tonnes dwt. These ships take their name from the fact that being too large to transit the Panama Canal they trade from the Atlantic to Japan via the Cape of Good Hope.

(c) *Panamax*, as their name implies are designed around the maximum dimensions allowed by the Panama Canal, these dimensions are length 274.3m beam 32.8m of the locks and draft 12m in the Lakes so Panamax size ships are around 79,000 tonnes dwt. The deadweight can vary depending on the shape of hull that the designer has chosen within the restrictions imposed by these dimensions. It should be noted that at the present time the 12m draft applies to the rainy season only, at other times of the year, the water level in the Lakes falls and ships may have to load to a reduced draft.

(d) *Handymax*. This is a development of the handysize bulk carrier and are ships of between 35,000 to 50,000 tonnes deadweight.

(e) *Handysize*. This is the smallest bulk carrier, as their name implies they are small ships of between 10,000 and 30,000 deadweight capable of carrying a wide range of different types of bulk cargo. They are mostly fitted with cargo handling equipment, the sophistication of which will depend on the type of cargo which the owner had in mind when considering what they wanted their ship to carry.

Plate 8 An example of a large "Capesize" bulk carrier, the *Mineral Nippon* (194,744 tonnes dwt), built in Japan at the Mitsui Engineering and Shipbuilding Co. Ltd.

COMBINATION CARRIERS

A problem that has exercised the minds of economists and ship operators has been that large bulk carriers, whether dry cargo or tanker, are usually unable to find a return cargo and therefore spend a large proportion, possibly up to half of their time in ballast. When a ship is said to be in "ballast" it means that she is only carrying seawater ballast to keep the ship in a seaworthy condition in order to make a passage to the next loading port, she is therefore not earning any freight while on this passage.

The idea behind the combination carrier is to design a ship that is able to carry liquid as well as dry cargoes at different times in order to avoid or reduce the unprofitable ballast voyages. In practice there are very few trading opportunities like this even when the voyage is extended to three or four stages. The alternative is for the combination carrier to be employed in an opportunistic manner, trading in oil cargoes when oil rates are higher than dry cargo rates and trading in dry cargo when freight rates are highest in this sector. There can be problems in changing from dry to liquid cargoes, as the remains of the dry cargo can obstruct tank valves and damage hatch seals allowing gas to escape, while the gas remaining after a liquid cargo has been discharged can be hazardous when handling dry cargoes, particularly where grabs are being used. It is for this reason that charterers prefer to use dedicated tonnage, ie dry bulk carriers for dry trades and tankers for oil trades, or failing this, combination carriers that have traded in one commodity, either wet or dry, for some time are preferred to ships that have just changed over. Many of the combination carriers in service today were built during the 1970s. As these are gradually replaced with newer tonnage it is possible that charterers will change their views over the employment of this type of ship.

There are two types of combination carriers.

The ore/bulk/oil carrier (OBO)

This is a ship with a number of large centre holds/tanks. She is constructed in such a way that these holds are used for whatever cargo the ship is designed to carry. These cargoes could be any one of a range of dry bulk cargoes or liquid oil cargoes. The large tanks/holds have to be cleaned out between the different cargoes and this can cause difficulties when changing from a wet to a dry cargo or vice versa.

The ore/oil carrier

This is a larger ship and, as its name implies, is designed to carry iron ore in a strengthened centre hold, or oil in side and bottom tanks and in some ships, the hold as well. These ships are larger than the OBO, with a deadweight of about 150,000 tonnes.

As the only dry cargo that these ships are likely to carry is iron ore, the centre holds are designed with this in mind and are therefore comparatively small, when taking the size of the ship into account (see Figure 2b).

SELF-DISCHARGING BULK CARRIERS

The smaller bulk carriers are almost always fitted with cargo handling equipment. This is because these ships are engaged in a wide variety of trades and they need the ability to load and discharge their cargo without having to rely on shore-based equipment. This type of equipment usually consists of cranes or derricks although some of these ships will be fitted with gantry cranes. When it comes to the larger size of ship—Panamax and above—this type of equipment is unable to cope with the quantities involved so these larger ships usually rely on shore equipment. There are, however, a few large bulk carriers with self-discharging equipment. The high cost of this equipment means that these ships are usually built with a particular trade or trades in mind and not employed on the charter market. The equipment is either conveyor or gravity belt systems and can produce very high discharge rates. Self discharging ships are commonly found in the Great Lakes where the short distances between the lake ports means that ships spend a higher proportion of their time in port than is usual in the deep sea trades and therefore the saving in port time due to the faster discharge rates more than compensates for the additional cost of the cargo handling equipment.

THE AGE OF THE BULK CARRIER FLEET

Table 74 shows today's bulk carrier fleet by type and year of build. From the table it can be seen, both by tonnage and number of ships, that more than half of the world fleet is over ten years old and a third is over 15 years old. This is important when considering the question of coverage insurance— see Chapter 9.

The practical problem is that if these older ships are to be replaced, freight rates must rise in order to cover the building costs. At present the market freight rates are too low to justify the newbuilding price of a bulk carrier.

There is also a shortage of facilities to scrap these ships, with a large number of tankers coming up to around 15 years old, there are likely to be more ships sold for scrap than there are facilities to scrap them over the next decade.

LOSSES OF BULK CARRIERS

One factor that has caused concern in recent years has been the high rate of loss and serious damage to bulk carriers. In 1990 23 ships were either seriously damaged or lost. This has promoted a research programme by

Table 74 Bulk carrier fleet including ore carriers as at 1 July 1991

tonnes	Up to 1971		1972–1976		1977–1981		1982–1986		1987–1991		Total	
	No.	D.W.	No.	D.W.	No.	D.W.	No.	D.W.	No.	D.W.	No.	D.W.
10–19,999	147	2,512,427	159	2,507,707	207	3,520,512	81	1,292,073	17	249,475	611	10,082,194
20–29,999	220	5,626,341	368	9,449,488	369	9,395,050	303	7,806,426	51	1,308,149	1,311	33,585,454
30–39,999	95	3,252,053	231	7,959,929	226	8,072,327	370	13,370,307	51	1,885,907	973	34,540,523
40–49,999	46	2,022,109	47	2,039,886	61	2,655,935	253	10,989,416	86	3,723,855	493	21,431,201
50–59,999	38	2,101,381	81	4,314,084	31	1,699,322	26	1,404,435	12	642,060	188	10,161,282
60–69,999	24	1,543,830	87	5,566,787	102	6,477,017	168	10,882,765	90	6,048,079	471	30,518,478
70–79,999	16	1,216,309	34	2,527,198	29	2,128,715	33	2,440,298	20	1,438,474	132	9,750,994
80–89,999	7	579,692	6	487,963	7	577,566	8	683,019	1	88,309	29	2,416,549
90–99,999	5	466,741	1	92,828	1	93,070	2	187,061	1	96,143	10	935,843
100–109,999	8	850,979	6	635,451	3	318,662	1	105,577	0	0	18	1,910,669
110–119,999	7	794,017	33	3,826,834	10	1,175,968	1	111,695	2	229,486	53	6,138,000
120–129,999	5	624,282	24	3,004,209	15	1,898,486	4	511,899	2	244,200	50	6,283,076
130–139,999	0	0	6	800,453	11	1,486,429	14	1,890,738	5	,682,121	36	4,859,741
140–149,999	5	712,762	7	1,009,368	10	1,420,943	20	2,881,354	24	3,549,215	66	9,573,642
150–159,999	4	617,742	7	1,099,030	0	0	6	909,957	14	2,125,473	31	4,752,202
160–169,999	4	656,924	7	1,150,009	1	169,749	10	1,657,613	3	505,637	25	4,139,932
170–179,999	0	0	1	178,750	0	0	15	2,612,632	4	704,037	20	3,495,419
180–199,999	0	0	1	190,903	3	583,198	14	2,692,350	11	2,040,959	29	5,507,410
200–219,999	0	0	1	218,359	0	0	5	1,033,873	13	2,688,057	19	3,940,289
220–239,999	0	0	1	227,557	0	0	3	680,732	3	688,159	7	1,596,448
240–259,999	0	0	0	0	0	0	2	510,779	1	245,609	3	756,388
260–279,999	0	0	3	818,825	0	0	1	267,889	1	260,000	5	1,346,714
280–299,999	1	282,462	0	0	0	0	0	0	0	0	1	282,462
Over 300,000	0	0	0	0	0	0	1	364,767	0	0	1	364,767
Total	632	23,860,051	1,111	48,105,618	1,086	41,672,949	1,341	65,287,655	412	29,443,404	4,582	208,369,677

Source: Simpson, Spence & Young.

103

Lloyd's Register of Shipping to try and ascertain the cause. The one common factor was the extreme age of the ships; the 23 ships that suffered serious damage or were lost in 1990 had an average age of 19 years while average age of the casualties over the previous ten years was 16. From figures published by Lloyd's Register the breakdown of the causes was:

- 70% side structure failure.
- 15% no information.
- 5% cargo shift.
- 10% miscellaneous.

Unfortunately many of those ships where there was no information were lost without trace.

The study has not yet been completed so many of the results must be conjecture but some of the factors that are currently considered to contribute to this are discussed below.

Corrosion

These ships have large areas of steel in their hull, and although the classification society surveys require the hull to be inspected for corrosion it is not easy to examine the whole area and there is ample evidence from other sources that these inspections do not detect all the damage.

If the paint coatings are allowed to break down then corrosion will be rapid in this area. This can easily occur inside a ballast tank without being noticed.

The sulphur content of some of the more high sulphur coals could contribute to the corrosion by producing sulphuric acid, which attacks the mild steel used in shipbuilding.

Physical damage to the vessel

The modern loading and discharging methods can put a considerable stress onto the hull. Heavy grabs used for discharging bulk cargoes can weigh up to 35 tonnes and if these are allowed to drop onto the tank tops or other parts of the ship's structure then considerable damage can result. Some of the smaller bulk carriers may have been used to carry steel coils.

The use of bulldozers to pile bulks into the centre of the hold is common practice. If these are allowed to strike the ship's side the resulting damage might lead to some of the side fracture failures. Pneumatic hammers may have to be used to loosen some cargoes that have become compacted before they can be discharged.

Allowing cargo to be loaded too fast, particularly into an empty hold where there is nothing to cushion the structure from the shock (which is also bad practice where coal is concerned because it leads to the release of methane in some coals), can seriously weaken the structure if repeated over a period of time.

Overloading

In a terminal where there are high loading rates right up to the end it is easy for the ship to be overloaded. If the loading is not stopped quickly enough when the maximum permitted draft is reached, she will be overloaded.

Many of the ships lost were carrying iron ore at the time and the high density of this cargo can put additional stress onto what was an already weakened hull. In the case of ships chartered for coal cargoes, if the ship is over 15 years old and has carried a large proportion of high density cargoes in the past, there is always the possibility that the hull has been weakened. In such a case the ship could suffer severe damage from what, in a newer ship would be considered a minor incident.

This problem of losses is certainly a cause for concern. It is likely that a large number of ships in service today are deficient in structural strength. The lessons learned from these unfortunate disasters can be fed into the design of the new ships but unless freight rates improve significantly there is not likely to be a major fleet replacement programme, owners being forced to continue with their old and possibly flawed ships because they cannot afford to replace them.

ENGINES FOR BULK CARRIERS

The diesel engine is the main form of propulsion for bulk carriers, although the steam engine was attractive a number of years ago because of its greater reliability and ease of maintenance. The higher fuel consumption of the steam ship when compared with the diesel engine meant that after the increases in oil prices in the 1970s the steam engine with oil fired boilers was no longer considered to be a realistic means of propulsion for bulk carriers.

A problem of ship operation with diesel engines is that of fuel quality. Marine fuel oils are of two main types; *diesel oil* is a distillate fuel, which as its name implies is suitable for diesel engines but it costs about twice as much as fuel oil. *Fuel oil* is also known as a "residual oil" and is obtained from the residues that remain after the crude oil has been refined and the more valuable fractions removed.

With the increases in oil prices, the oil companies have endeavoured to extract more of these fractions from the crude oil with the result that, as refining technology has improved, there has been a corresponding decrease in the quality of the residual oils.

One of the reasons that the steam engine retained its popularity for so long is that the marine boiler is able to burn a wide range of different grades of fuel and hence the declining quality of the marine fuels did not present a problem for these engines.

This posed a challenge to the makers of diesel engines. The early diesels all ran on diesel oil, but in the 1960s there was a gradual change to fuel oil. As a result of the changeover, the engines suffered increased wear and breakdowns when they were first run on fuel oil, but improved design, cylinder lubrication and fuel oil treatment have largely eliminated these problems and today the diesel engine is almost as reliable as the steam engine.

The reduced maintenance requirements have led to reductions in the number of engineers carried in a typical bulk carrier from around seven in the 1960s to three on some modern ships.

The problems that remain over fuel quality are frequently caused by a ship being forced to take bunkers of unknown quality. There are recognised quality standards for bunker fuel but it is not always possible to get bunkers that meet this standard. Thus engine damage may result from the use of bunkers whose exact specification remains unknown. Another problem is to do with compatibility. Fuel oils from certain different sources should not be mixed as there is a risk of sludge formation if they are incompatible. Some fuels have a high pour point (the temperature at which they can be pumped) so they tend to solidify in ships' tanks and cannot be used or even pumped. The only way to prevent this is to maintain the fuel at a temperature 10°C above the pour point by steam heating. These problems can severely restrict an owner's or charterer's ability to take advantage of cheap bunkers.

Today almost all bulk carriers are powered by one or other of the two main types of diesel engines, the slow or medium-speed diesel.

The problem faced by engine designers and naval architects is that for maximum efficiency the marine propeller should turn at as slow a speed as possible. As the speed increases, the propeller "cavitates" which means that air bubbles are produced on its face and it loses efficiency. The main constraint on propeller speed is the size of the ship's stern frame—the slower the propeller turns the larger it needs to be in order to produce the required power until the point is reached where it can no longer be accommodated in the stern frame without an unacceptable increase in the ship's draft. Thus, in practice, the slowest propeller speed that can be achieved in a Panamax size of ship is about 80 rpm.

The slow speed engine is a large two-stroke engine, which turns at speeds of between 80 and 120 rpm and is coupled directly to the propeller without the need for an expensive gearbox. Reversing the propeller to stop or turn the ship is achieved by reversing the engine. This engine has the advantage of being able to burn the same grades of fuel oil as the steam engine, but in order to produce the required power at the very slow speeds required by the direct drive to the propeller, it has to be very heavy and requires a large amount of space. These engines can have five to 12 cylinders. With the large bore that is common today, this results in engines of about 18–20 m long and a weight of about 300 tonnes. This can reduce the amount of space available for cargo. The higher weight can result in a reduced deadweight available for cargo in those bulk carriers that are designed to carry high density cargoes. The twin advantages of low maintenance and good reliability outweigh these disadvantages and this type of engine is found on most bulk carriers.

The medium-speed engine is a smaller unit turning at between 400 and 600 rpm and can be two- or four-stroke. For a ship fitted with medium-speed diesel engines, a gearbox is required in order to bring the engine shaft speed down to the 100 rpm that is the most practical speed for a propeller. These engines are usually unidirectional with the gearbox being used to reverse the propeller or in some cases a controllable pitch propeller. On some ships the medium speed engines are fitted in pairs with a common gearbox. Although these engines are more economical on fuel than the slow-speed diesel, they require more maintenance and low grades of fuel can cause problems. In some cases it is necessary to blend diesel oil with the fuel oil if the engine is to run efficiently. The main advantage with these engines is that they are lighter and more economical in space than the slow-speed diesel.

Boilers

On a bulk carrier there is little requirement for steam, the main purpose is for heating the fuel and for domestic heat. With a combination carrier, however, there is the further requirement for powering cargo pumps and the possible need to heat cargo, so these ships have a greater requirement for steam than the pure bulk carrier. To reduce fuel consumption when at sea, the diesel engine exhaust gasses are passed through a heat exchanger and used as the source of heat to raise steam. This is known as an exhaust gas boiler or an economiser, and can provide enough steam to meet the needs of the bulk carrier when at sea. In most ships these boilers are also provided with oil burning arrangements for use in port when the engine is not running. The combination carrier will be provided with a larger capacity boiler in order to provide enough steam to drive the cargo pumps.

Electricity generation

The traditional method of generating electricity was, and still is on many ships, by using diesel alternators. These suffer from the disadvantage of having to use diesel oil and although the consumption is only about one or two tonnes per day, the price of diesel oil is about twice that of fuel oil and this can therefore prove expensive when taken over the life of the ship. There is also the additional maintenance requirement to be considered. On newer ships the electricity is generated when at sea from the main engine by means of a shaft alternator. This is driven by a take-off from the main engine, thus giving the advantages of reduced maintenance and the use of the cheaper fuel oil instead of diesel for electrical generation. The diesel alternator is still carried for use in port and when the main engine is not running at the required steady speed for power generation.

Coal as a fuel for bulk carriers

There were two fuel price shocks for shipping in the 1970s. The first was when the cost of crude oil was quadrupled by the Organisation of Petroleum Exporting Countries (OPEC) in 1973 and the second when the price of oil was trebled in 1979 in the wake of the Iranian revolution. This led to a number of designs for coal-fuelled ships. The traditional engine for the bulk carrier is the slow-speed diesel engine. This is a reliable power plant that, when fitted in a modern hull design, can prove very economical, but when fuel oil prices were around $200 per tonne, the use of coal-fired steam turbine ships appeared attractive. The Australians went as far as to build two ships for carrying coal to Japan but since that time fuel oil prices have fallen and with fuel oil costing around $60 per tonne this type of propulsion is no longer economic.

About the same time a number of experiments with sail assisted bulk carriers were carried out. The Japanese built a number of ships, but the fall in prices of fuel oil caused interest in these developments to cease.

Table 75 Engine powers for different types of bulk carrier

Ship Type	Dwt	Speed	Engine	Horsepower	Consumption
150,000 +	162,437	16	12 Cyl B & W	28,100	101
Capesize	119,510	15	8 Cyl B & W	24,800	82.5
Panamax	60,784	15	8 Cyl Sulzer	16,000	53
Handymax	49,751	14	12 Cyl M.A.N	12,000	40 *
Handysize	24,355	14.5	6 Cyl Sulzer	8,910	33.4

* Medium speed diesel
NB: The speeds quoted in the reference books are usually those obtained from trial data with a new ship in ballast condition. As the ship ages, her hull gets rougher. This will cause a reduction in speed, in addition, the loaded service speed is usually one knot slower than the speed in ballast condition, thus the information regarding speed should be treated with caution.

Fuel consumption is in tonnes per day and does not take any diesel oil for the alternators into consideration.

THE RELATIONSHIP BETWEEN SPEED AND CONSUMPTION

The voyage cost of a bulk carrier mainly comprises the fuel costs. The other costs are the port costs, which are not very large as the cargo handling is paid by the charterer. The consequence of this is that when freight rates are low and fuel prices high, the use of the most economical speed for a particular voyage can make the difference between profit and loss.

Owners should be aware of the fuel consumption for their ships over the range of operational speeds. These data should be obtained during the acceptance trials and verified in service by using a flow meter to measure accurately the consumption under a range of different service conditions, both loaded and in ballast. Apart from its use in voyage estimating, these data are of value in monitoring the ship's performance and enabling remedial action to be taken if it is found to deteriorate beyond what the owner considers to be acceptable limits. Apart from the financial benefit to the owner, this information is of value when the ship is on time charter and owners want to protect themselves against any claims of underperformance that the charterer may make.

In many cases these trials have not been carried out and it is necessary for the owner to resort to more empirical methods. There are two methods of calculating fuel consumption over a range of speeds given the consumption at one particular speed. The first method, which has been used for many years, is the so called "cube law".
The formula used is:

$$\text{New Consumption} = \frac{(\text{New Speed})^3}{(\text{Old Speed})^3} \times \text{Old consumption}$$

This is well understood but it is not as accurate as the more recent "J" curve. This was developed by Professor P M Alderton while he was a lecturer at the City of London Polytechnic. The method used for calculating the consumption is to multiply the consumption at 14 knots by a series of multipliers or "J" factors, these are obtained by:

$$J = AV^3 - BV^2 + CV$$

Where V is the speed at which the consumption is required and A, B and C are constants.

Approximate values of the constants are:

$$A = 0.000459$$
$$B = 0.0048$$
$$C = 0.04868$$

More accurate values for these constants relating to a particular ship can be obtained from direct measurements of the consumption, possibly during trials.

Using the above formula the following table of values for "J" has been calculated, using the consumption from the Handymax in Table 75 as an example. The fuel consumptions at the different speeds are calculated by multiplying the consumption at 14 knots by the "J" factor using the following formula:

New Consumption = J × Consumption at 14 knots

Table 76 Table of values for "J" factor

Speed	"J"Multiplier	Consumption
10	0.47	18.8
11	0.57	22.8
12	0.69	27.6
13	0.83	33.2
14	1.0	40
15	1.2	48
16	1.43	57.2

NB: the values of the consumption are quoted to one decimal place to enable the reader to follow the calculation, when this calculation is being used for operational purposes the resulting consumptions should only be relied on as being accurate to the nearest ton per day.

The important factor that emerges from the calculation is that a change in the ship's speed of only one or two knots can make a substantial difference to the fuel consumption.

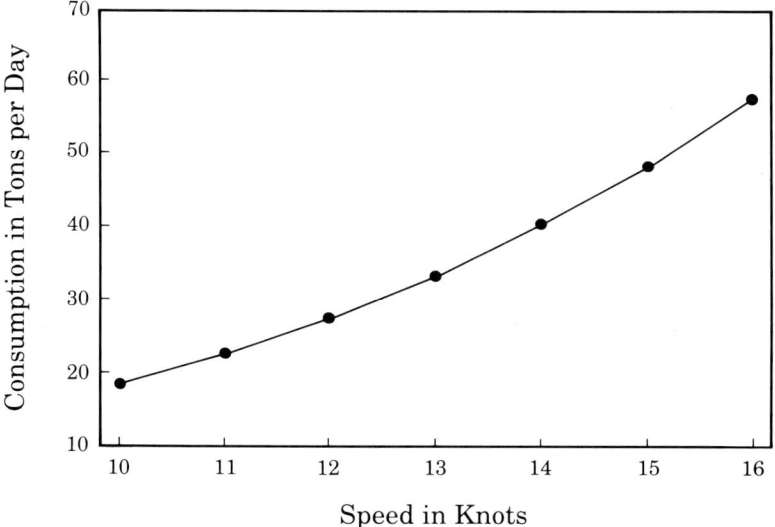

Figure 3 Relationship between speed and consumption

The graph in Figure 6 shows the relationship between speed and consumption. It should be noted that in this range the relationship is a gentle curve but once the speed increases to about 20 knots, the increase in consumption is very rapid. Such a speed would not be practicable for a bulk carrier so the calculations have only been carried out over a range of speeds that would be realistic for this type of ship.

A situation where an owner can take advantage of the reduced consumption at slower speeds is where it is known in advance that a ship will have to wait at anchor outside a port until the berth is available. Should it be possible to book the berth in advance and adjust the ship's speed in order to arrive when the berth is free, the owner or charterer could save a considerable amount of money.

This calculation can be applied when working out a voyage estimate, when freight rates are low and fuel costs relatively high. The adjustment of the ships' speed by a knot or two can make a difference to the cash flow for a given voyage.

Least cost speed

Using the 150,000 + bulker in Table 75 over an imaginary voyage from Richards Bay in South Africa to Yokohama, a distance of 7,481 miles, the following results were obtained.

Taking the given fuel consumption of 101 tonnes per day at 16 knots, we get the fuel consumptions shown in Table 77.

Table 77 Fuel consumption

Speed	"J"	Consumption
10	0.47	33.2
11	0.57	40.2
12	0.69	48.7
13	0.83	58.6
14	1.0	70.6
15	1.2	84.7
16	1.43	101

Using a spreadsheet to speed up the calculations, and assuming the cost of fuel to US$100 per ton and a daily running cost, or time charter cost of US$10,000 per day, the following results were obtained.

The consumptions at the different speeds were taken from Table 77, and the following calculations were made for each of the speeds under consideration.

Number of days for the passage (Days)
Total fuel required for the passage (Fuel)
Total cost of the fuel required for the passage (F Cost)
Time charter cost for making the passage (O Cost)

Total cost which is the sum of the fuel cost and the time charter cost (T Cost)

Table 78 Cost/speed

Speed	Con	Days	Fuel	F Cost	O Cost	T Cost
10	33.2	31.2	1035	$103,487	$311,708	$415,196
11	40.2	28.3	1139	$113,915	$283,371	$397,286
12	48.7	26.0	1265	$12,650	$259,757	$386,259
13	58.6	24.0	1405	$140,509	$239,776	$380,284
14	70.6	22.3	1572	$157,190	$222,649	$379,839
15	84.7	20.8	1760	$176,011	$207,806	$383,810
16	101	19.5	1968	$196,766	$194,818	$391,584

The table shows clearly that the least cost speed is around 14 knots, considerably less than the maximum possible speed.

Figure 4 demonstrates this relationship:

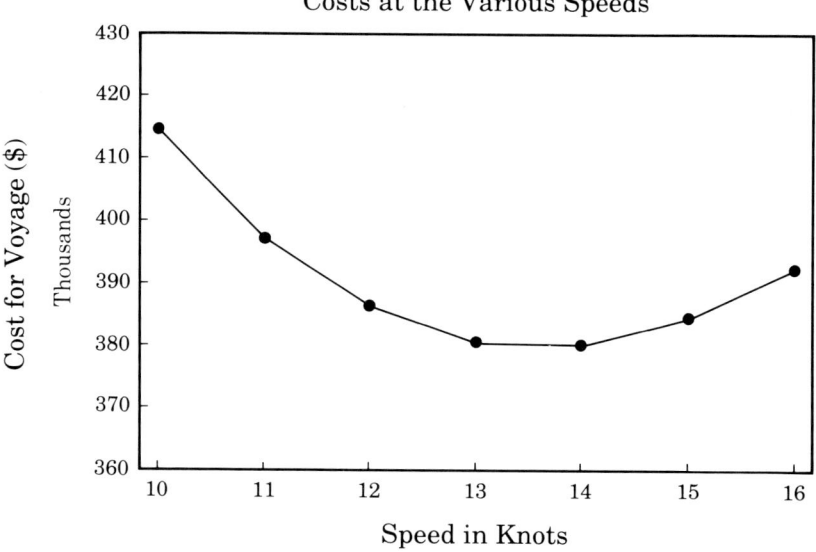

Figure 4 Voyage from Richards Bay to Yokohama.

From the graph in Figure 4 it can be seen that the owner/time charterer can accept a speed of between 13 and 14 knots for the voyage, but should not agree to any speed greater than 15 knots or less than 13 knots. The reader can try the calculation using different costs to check on the result, but in general the greater the fuel cost the lower the speed and the greater the time charter cost the higher the speed.

The least cost speed is the speed that a time charterer would use as it minimises the cost of the voyage. In this calculation there is no attempt to include inventory cost; with a cargo of coal this cost is unlikely to be significant.

Most profitable speed

The shipowner on a voyage charter is interested in profit per day and therefore in reducing the number of days that the voyage takes. This speed will be higher than the least cost speed when the voyage is profitable, equal to the least cost speed when the profit is zero, and less than the least cost speed when the voyage is run at a loss. In order to calculate the most profitable speed, the calculation is extended to include the daily profit at the different speeds. In general fuel costs have to be very high, or freight rates very low before it is to an owner's advantage to steam at slow speeds.

Practical considerations

These calculations do not take any limitations of the ship's engine layout into account. On ships fitted with a shaft alternator it is frequently the case that it will only operate over a narrow speed range. Should the ship steam at a reduced speed it may be necessary to run a diesel alternator if it is not possible to use the shaft generator. In such a situation it is unlikely that there will be any financial advantage in reducing the speed.

In order to steam at a reduced speed it may be necessary to alter the fuel injection timing and fit slow-speed injector nozzles if the engine is to run efficiently.

TONNAGE

How big is a ship? This is in fact a very difficult question to answer as the ton when used to describe the size of a ship can have many different meanings.

The word "ton" when used in relation to a ship's carrying capacity was originally spelt "tun" and referred to the container (similar to a barrel) of 252 gallons used in the wine trade. In the medieval ships the tonnage of a ship literally meant the number of these tuns that a ship could carry and so was a very useful measure of a ship's earning capacity. Since then it has become more complicated as different forms of measurement have been introduced and as they are used to assess the various 'dues' or charges that may be levied on the ship, a number of attempts have been made to minimise the various tonnages while at the same time maximising the ship's cargo carrying capacity.

IMO tonnage

In July 1982 a new international system of measurement for ships came into force. The word "ton" which in fact, consisted of 100 cubic feet, because of the confusion entailed, is no longer to be used as a unit of measurement.

Under the new rules *gross* tonnage is the volume of all the enclosed spaces in the ship measured in cubic metres and multiplied by a factor (K), which varies according to the type of ship.

Gross tonnage is measured by:

$$GT = K_1V$$

where V is the total volume of all enclosed spaces in cubic metres.

For a bulk carrier, $K_1 = .2 + .2Log_{10}V$.

Gross tonnage is a measure of the overall "size" of a ship, therefore most safety regulations are written around this figure as in general, the larger ship is, the higher the standards of safety equipment that the ship is required to carry.

Net tonnage is a measure of the ship's cargo space and therefore her earning capacity so dock harbour and other dues are levied on the net or register tonnage.

Net tonnage is measured by:

$$NT = K_2V_c(4/3 \times d/D)^2$$

where $K_2 = .2 + .2log_{10}V_c$
   D = Depth of the ship
   d = Summer draft of the ship
   V_c = Volume of the cargo space in cubic metres

National tonnages

Prior to the introduction of the IMO measurements, ships were measured for tonnage according to the rules of their country of registry. Ships built before 1982 have until 1994 to change over to the new system and as there are advantages for some ships to remain measured under the old rules, both methods of measurement will be in force until then.

Gross tonnage

This is the volume of the enclosed space in cubic metres and was converted at 2.83 cubic metres to the ton (100 cubic feet).

Net or register tonnage

This is gross tonnage after the deductions for crew accommodation, bridge, engine room and other spaces that do not contribute to the ship's earnings have been made. Thus "net" is a measure of those spaces concerned with the carrying of the cargo.

Suez and Panama Canal tonnage

Both these authorities have their own rules for the measurement of gross and nett tonnage and ships using the canals are charged on these tonnages. Suez tonnage is based on the Danube rules.

Grain and bale space

Because of the complicated rules of tonnage measurement, net tonnage is not an accurate measurement of a ship's cargo space. Instead the capacities of the spaces in cubic metres (or feet) are quoted. Two measurements are given—grain and bale space. The grain space is the larger of the two as loose grain will flow into spaces that could not accommodate a more bulky cargo.

Thus for a cargo of coal the larger space is used, although with most of the bulk carriers used in the coal trade, the holds are clear of any obstruction and therefore there is only one space quoted, this is the space available for the cargo.

Bill of lading (B/L) tonnes

This is the tonnage of cargo loaded as stated on the bill of lading and it is on this figure that the freight is paid. This will be the figure obtained in the loading port either by "draft survey" or by weighing (see reference in next chapter).

Displacement

From the principal of Archimedes, a ship floats, when the underwater portion of the hull displaces a weight of water equal to the weight of the ship. Thus by calculating the underwater volume of the ship and multiplying it by the density of salt water (which is taken to be 1.025) the total weight of the ship and cargo, in the case of a loaded ship, can be obtained. This is the basis of the draft survey described in the next chapter.

Light displacement

This is the weight of the empty ship, ie a ship with no cargo, stores or fuel on board but including those spare parts that she is required to carry by law.

Load displacement

This is the weight of the ship and contents including fuel, stores and cargo when she is loaded down to her summer loadline.

Deadweight

This is the difference between the light and load displacement and represents the weight that the ship can carry. Do not confuse deadweight with cargo capacity. Deadweight consists of all the weight the ship can carry, which includes fuel and stores. Shipbrokers frequently use the expression deadweight cargo capacity (DWCC) to mean the cargo that the ship can load.

Table 79 Examples of different ships

Tonnage	Handysize	Handymax	Panamax	Capesize	150,000+
Nett	7,730	13,689	30,410	44,545	96,331
Gross	11,931	19,264	36,785	65,135	120,787
Deadweight	19,126	36,935	64,103	119,500	214,592
Suez	10,284	17,931	32,888	59,353	111,485
Panama	9,017	17,316	32,709		

Paragraph ships

Most shipping regulations are based on gross or deadweight tonnage. Owners therefore build ships to take advantage of the different regulations, hence ships of 499 and 1,599 gross tonnages are very popular. Such ships are known as "paragraph ships" because they take advantage of a paragraph of the regulations to avoid having to comply with the more stringent safety regulations that apply to the larger size of ship.

For example, a ship of 1,600 gross tonnage or more is required to carry a radio operator. Thus by restricting the gross to 1,599 and maximising the deadweight, the owner is able to obtain the largest possible ship without having to carry the extra crew member.

REFERENCES FOR CHAPTER 6

W.D. Ewart, *Bulk Carriers* (Pub Fairplay).
P.M. Alderton, *Sea Transport Operation and Economics* (This gives a more detailed treatment of the most profitable speed problem)

CHAPTER 7

The loading, carrying and discharging of coal

There are a number of considerations that must be taken into account
before a cargo can be loaded, the most important of which is to calculate
the amount of cargo that the ship can carry after taking the various
loadline zones into account. This requires information about the loadline
zones and the various deadweights that the ship can load to.

LOADLINES

Before they are allowed to proceed to sea, all ships are required to be
surveyed and marked with a loadline. Governments frequently delegate
the function of surveying and assigning the loadline to an assigning
authority such as Lloyd's Register. The assigning authority will decide on
the ship's summer freeboard which is distance between the deckline
and the top of the line through the loadline disc (see Figure 5).

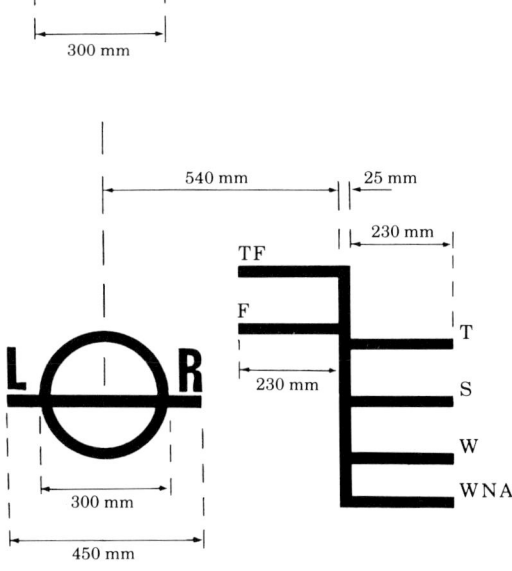

Figure 5 Loadline.

The tropical and winter lines are then placed 1/48th of the ship's summer draft above and below the summer mark.

The fresh water lines are placed above the summer line so that a ship floating to this line in fresh water would rise to her summer or tropical lines when sailing from fresh to salt water.

The winter North Atlantic mark only appears on ships less than 100m in length and is placed 50mm below the winter line.

To calculate the deadweights for a ship the tonnes per centimetre immersion (TPC) or tons per inch immersion (TPI) must be obtained from the ship's particulars. The value of 1/48th of the summer draft is obtained and multiplied by the TPC or the TPI in order to obtain the tonnage differences.

Note that ships are measured in both metric and imperial units. It is important therefore that anyone concerned with ship operations is familiar with both systems.

Example

A bulk carrier has a summer deadweight of 29,020 tons. At this deadweight her summer draft is 35.34 ft and her TPI at that draft is 96.4 tons. To find her tropical and winter deadweights and drafts, take 1/48th of the summer draft:

Correction to summer draft is = 35.34 /48 = 0.7425 ft = 8.9 ins. To find the tropical and winter deadweights, the draft difference is multiplied by the TPI = 8.9 × 96.4 = 858.9 tons, rounding to 859 tons.

Therefore tropical deadweight = 29,020 + 859 = 29,879 tons and the corresponding draft is 35.34 + 0.7425 = 36.08 ft.

The winter deadweight is = 29,020 − 859 = 28,161 tons and the corresponding winter draft is 35.34 − 0.7425 = 34.60

Loadline chart

The loadline chart (see Figure 6) indicates the different loadline zones. Some are permanent and some are seasonal. A ship is required under the International Loadline Rules to be loaded to her correct loadline when she enters a zone at sea as well as when proceeding to sea after loading a cargo.

Boundary ports are ports on the boundary between two zones. A ship arriving at a boundary port must be loaded for the zone that she passes through on the way to the port. When leaving port the ship must be loaded for the zone that she will enter after sailing.

Loading calculations

In order for a ship to be correctly loaded at all stages of the voyage it is

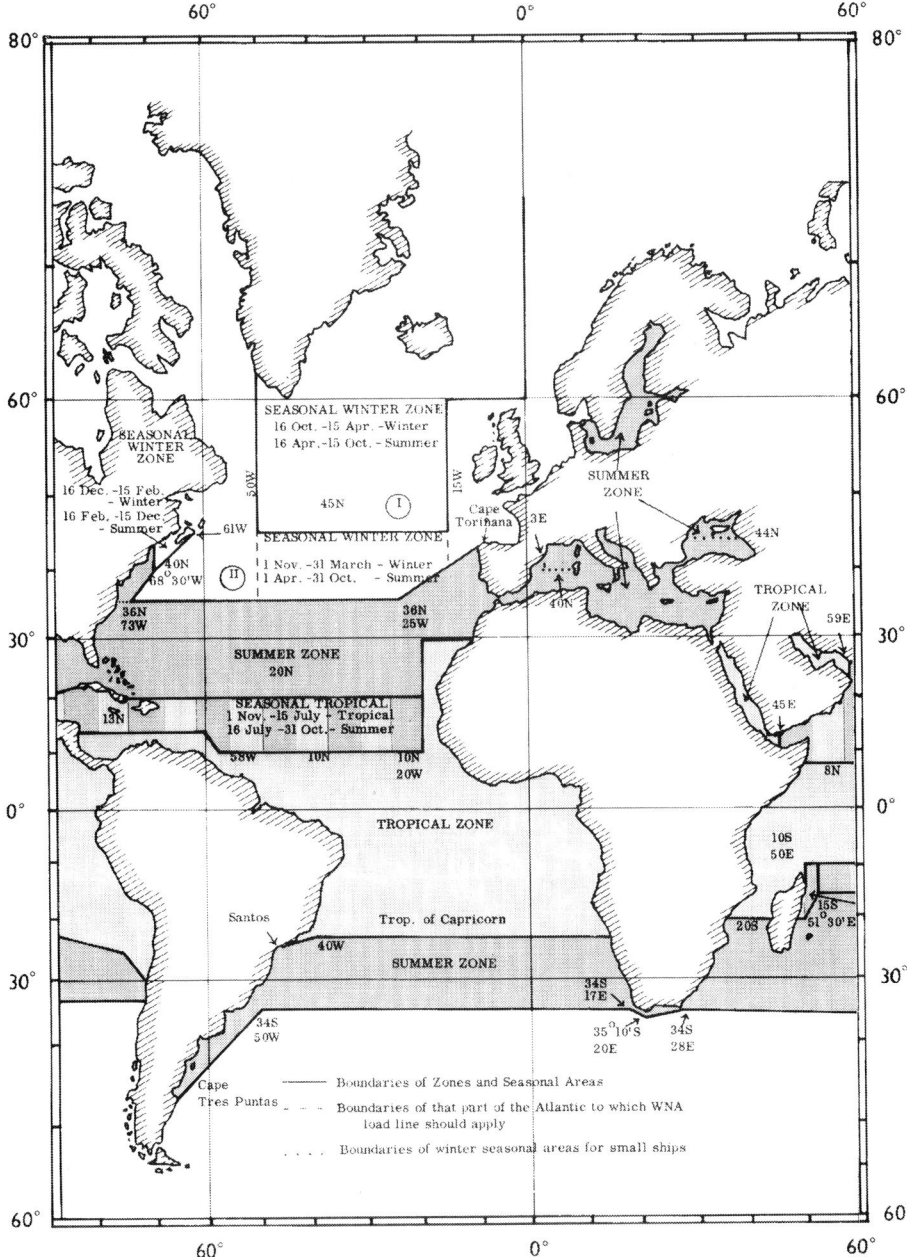

Figure 6 Loadline chart.

necessary to calculate the tonnages that she can lift at different stages of the voyage. One such method of calculating the tonnages is given below.

Example

A bulk carrier with the following particulars is to load a full cargo of coal at Richards Bay (South Africa) for Rotterdam. She is due to arrive at the loading port at the end of October. The particulars of the ship are:

> Summer deadweight — 103,100 tons
> Summer draft — 49.46 ft
> TPI at the summer draft — 225 tons
> Ship's speed — 14.5 knots
> Daily Fuel Consumption — 170 tons
> Water, stores and constant — 1,000 tons
> Allow five days' reserve of fuel throughout the voyage

The constant is the difference between the theoretical cargo that a ship can load and the actual amount of cargo as measured from her draft marks. In practice this constant increases as the ship gets older and can be considered to represent the spare parts carried as well as the paint and rubbish that accumulates on a ship.

Note that as it is not possible to forecast what problems or delays might arise during a voyage it is usual to calculate the amount of fuel required for a voyage and then add an additional amount for a reserve. In this example a reserve of five days is used.

By examining the loadline chart (Figure 9) it can be seen that it will be winter when the ship reaches Cape Finisterre, off northern Spain so the ship will be required to be on her winter loadline when she reaches this point. The distances can be taken from any set of distance tables, and they will vary depending on the tables used.

> Distance from Richards Bay to Cape Finisterre (N Spain) = 6,181 miles.
> Distance from Cape Finisterre to Rotterdam = 802 miles.

The first step with this problem is to calculate the ship's winter deadweight, as explained above on page 118.

Summer draft is 49.46 ft, dividing by 48 gives a reduction of 1.03 ft or 12.37 ins.

The deadweight reduction is the draft difference multiplied by the TPI (tons per inch immersion) = 12.37 × 225 = 2,782 tons. Hence the winter deadweight is 103,100 − 2,782 = 100,318 tons. She must, therefore, have a deadweight of 100,318 tons when she reaches Cape Finisterre.

The deadweight does not only mean the cargo but includes the fuel, water stores and any constant. These must therefore be deducted from the deadweight in order to obtain the cargo that the ship can lift.

The next step, therefore, is to calculate the fuel required for the voyage, not forgetting any reserves that are considered necessary.

The distance from Richards Bay to Rotterdam is 6,938 miles. At 14.5 knots (1 knot is a nautical mile per hour, this is slightly faster than a statute mile per hour) this will take 478.5 hours or 20 days (to the nearest day). The fuel required is therefore:

$$20 \times 70 + 5 \text{ (reserve)} \times 70 = 1,750 \text{ tons}$$

From Richards Bay to Cape Finisterre is 6,181 miles and the voyage will take 18 days. This will require 18×70 tons of fuel = 1,260 tons.

Therefore the maximum deadweight leaving Richards Bay will be 100,318 tons (winter deadweight that she must be at Cape Finisterre) + 1,260 (the fuel she will use on the passage from Richards Bay to Cape Finisterre) = 101,578 tons.

Hence in order that she is not overloaded when entering the winter zone off Cape Finisterre, she must leave Richards Bay with a deadweight of no more than 101,578 tons.

This statement begs the obvious question, how will anyone in authority know what the ship's draft was at that time when she was out at sea and far from the prying eyes of any inspector? It is an easy matter for an inspector in Rotterdam to read the draft of the ship on arrival and after allowing for the fuel consumed between Rotterdam and Cape Finisterre discover whether she was overloaded or not.

To complete the calculation, when leaving Richards Bay she will have on board 1,000 tons of water, stores, etc and 1,750 tons of fuel. Hence she can load 98,828 tons of coal (101,578 − (1,000 + 1,750)).

This calculation has been deliberately simplified to show the main points that have to be taken into consideration when deciding how much cargo a ship can load. Unless the details of the voyage and cargo were known before the ship bunkered it is unlikely that the master would be able to ensure that he had exactly the right amount of fuel on board. Also a ship like this is likely to use diesel oil for her generators but as the consumption is only in the order of one or two tons per day this would not affect the calculation.

STRENGTH OF SHIPS

After calculating the amount of cargo the ship can carry, the next problem is to decide where the cargo can be loaded.

With a large bulk carrier it is important that she is loaded in such a manner as to keep the hull stresses to a minimum. There have been a number of sinkings of large bulk carriers in recent years, the exact cause of these is unknown. It is not even known if there is a common factor other than the age of the ships concerned. The repeated loading and discharging of these ships over a number of years may well have lead to fatigue cracks in the ship's structure and these may eventually lead to a structural failure if a the hull is subject to a sudden stress.

A bulk carrier is subject to stress all the time she is afloat. The stresses can be divided into *local*, which only affect a small part of the ship's structure, and the *overall* stress, that affects the whole ship. There are two main types of overall stress, which affect bulk carriers. The first is *bending moment*, the extreme case of this is where a ship is supported at each end by a wave with a trough amidships. The resulting forces will cause the ship to sag putting the keel under tension and the deck under compression.

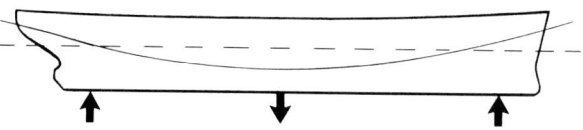

Figure 7 Sagging.

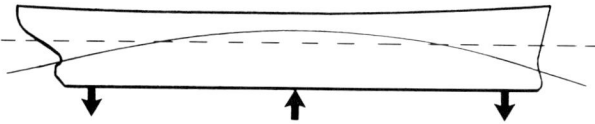

Figure 8 Hogging.

When the ship is supported amidships on the crest of a wave and with a trough at each end, she is said to be hogged. This puts the deck under tension and the keel under compression.

When a ship is in a seaway she will be subject to alternate sagging and hogging and if the weight distribution is not within acceptable limits this can fracture the hull. When a ship is at sea she is subject to greater stress than when in harbour. This means that ships have two sets of allowable stress, one for harbour and one for when the ship is at sea. This is further complicated by the cargo distribution and draft. When the ship is empty, there will be the weight of the engines at the stern and the fuel in the bow while the centre section is empty. This produces the situation where the weight at the ends and the buoyancy pushing up the centre section will cause the ship to hog. With a cargo loaded in the centre section she will sag. Thus when a ship is in port loading cargo, the stresses will change from hogging to sagging during the loading process. It is therefore vital that the cargo is loaded to a prearranged plan in order to prevent damage.

When checking the loadline to ensure that the ship is not overloaded, the procedure is to measure the distance from the deck line to the water on each side amidships. By hogging the ship, the amidships area and the loadline is raised with respect to the bow and stern. This will allow more cargo to be loaded, so in order to load the maximum cargo some chief officers will load the end holds first to hog the ship and so increase the amount of cargo carried. Provided the maximum allowed harbour stresses are not exceeded there is nothing wrong with this practice but it does raise some questions as far as older ships are concerned.

The second type of stress is that between compartments known as *sheer force*. Sheer force is the resultant force at the bulkheads that results from different forces in adjacent compartments. When a compartment is empty, the buoyancy of the compartment is acting upward. When the same compartment is loaded, the force of buoyancy is less or there could even be a resultant force downwards if the weight is greater than the buoyancy. This will naturally impose a stress at the bulkheads dividing these compartments. These stresses will again change as the loading or discharging operation progresses.

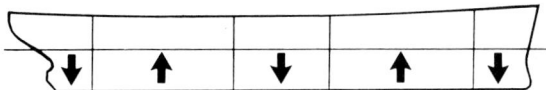

Figure 9 Sheer forces.

When the ship is designed, the maximum allowable stresses for both sea and harbour conditions are calculated. The ship is then supplied with a calculator or, with newer ships, a computer so that the stress levels associated with any stage of the cargo handling can be calculated and any loading or discharging programme can be checked to ensure that it does not exceed the maximum allowable stresses.

The local stresses do not pose a problem as far as the carriage of coal is concerned. The density of coal is not sufficiently high to stress the hull as is the case with iron ore. Ships that carry coal may also have carried some of the high density cargoes in the past so there is always the possibility that the structure may have been weakened by the incorrect stowage of a high density cargo such as iron ore (See the discussion of these losses in Chapter 7.)

CARGO HANDLING SYSTEMS

The dramatic increase in the size of bulk carriers over the last 20 years has only been possible because loading and discharge rates have increased in a similar manner. In fact the use of bulk carriers instead of tweendeckers in the coal trades has led to a reduction in port times.

Unlike containers, when bulk cargoes such as coal are taken into account different techniques for loading and discharging have to be considered. Many of the coal ports, such as Richards Bay in South Africa, only cater for loading ships, while other ports will be designed for discharging the large bulk carriers and possibly reloading some coal into small ships or coasters. Regardless of whether a ship is being loaded or discharged, the important principle is that as far as possible the cargo handling process should continue without interruption and time lost when changing between holds must be minimised. This is one of the areas where the requirements for fast loading or discharging might conflict with the requirements for reducing stress in the ship, since in order to keep stress to a minimum it is necessary to load the holds as evenly as possible. This will require more shifting between the holds than the terminal may want.

Cargo handling equipment

Cargo handling equipment for coal cargoes can be divided into two main types: grabs and continuous handling equipment.

Grabs

A grab is a constant weight device. It is possible to change the grabs depending on the type of cargo being handled so that the grab always picks up its designed weight. This will ensure that the crane is always working to maximum capacity. When discharging, the grab drops the cargo into hoppers situated under the crane. Each hopper can be designed to weigh the cargo as it comes out of the ship, releasing it onto a conveyor when the preset weight is reached.

Continuous handling equipment

The capacity of this type of equipment is based on volume. Thus a higher handling rate is achieved with high density cargoes such as ore than with low density cargoes such as grain. Coal having a medium density is ideally suited to this type of equipment.

Loading is usually carried out using loading arms on the jetty, fed by conveyors. These arms can slew to the different holds and rates over 1,000 tonnes per hour can be achieved from each arm. The loading arrangements can also be linked with blenders to ensure that cargo to the right specification can be loaded.

Plate 9 Grabs being used to discharge a cargo of coal at Rotterdam.

Discharging is by a range of equipment, a bucket wheel will lift the coal onto a conveyor or a vertical conveyor system can be placed in the hold and the cargo lifted out by means of an endless chain of buckets, again depositing the coal onto a conveyor.

Floating equipment

With some ports the ship may be loaded or discharged into barges when at anchor. In order to maintain the discharge rate the same type of handling equipment is used but floated out to the ships. This idea can be developed with the use of self-discharging barges—in the case of Canada, coastal ships discharging directly into the bulk carrier.

Regardless of how the cargo is to be discharged, there will always be residues remaining in the holds, and small bulldozers may be used to scrape the residues into piles so that they can be more readily picked up by the grabs. When there are pressures of time on the berth it is likely that the ship will sail with some residues remaining on board.

HAZARDS ASSOCIATED WITH THE CARRIAGE OF COAL

The carriage of coal by sea involves four possible hazards, the severity of which depends on the type of coal carried.

The first hazard is the emission of methane. The danger of explosions in coal mines is well documented, the dreaded "fire damp" and the use of the safety lamp in coal mines are well known. Coal cargoes may give off methane in different amounts, a mixture containing between 5% and 15% methane in air will explode when ignited. Methane is lighter than air and will therefore accumulate in the top spaces of the holds and possibly escape through the hatch covers. It is thus important that no "hot" work is carried out near a hold containing coal.

The second danger from carrying coal is that of spontaneous heating. This can occur in the centre of a cargo of coal and once the cargo has been loaded there is little or nothing that can be done to prevent it from heating. If this is allowed to continue it can lead to spontaneous combustion. It is therefore important that before loading a cargo of coal the temperature is checked to ensure that it is within the acceptable limits for the carriage of that particular type of coal. If the grade of coal carried is liable to spontaneous heating then the hold should be fitted with temperature sensors. This situation can be aggravated by the presence of oxygen in the cargo and so it is also important that no air is allowed to circulate through the hold. In addition the coal must be trimmed level after loading to reduce the surface area exposed to oxygen. Any moisture in the coal will encourage this heating.

The third hazard is that of the sulphur in the coal causing corrosion to the ship's structure. It is here that damp coal could prove to be an additional hazard as the moisture could turn to sulphuric acid.

The last hazard is the possibility of liquid movement. If the coal is in the form of fine particles or a slurry then there is a danger of liquid movement during the voyage. On a small ship this could lead to a loss of stability and a possible capsize.

The Code of Safe Practice for Solid Bulk Cargoes lays down a number of requirements for the carriage of different types of coal.

Among the general requirements there are a number of requirements for all coal cargoes, the more important elements of which are:

1. All spaces should be clean and dry, any residues from previous cargoes and any portable battens must be removed.

2. All electrical cables and components in cargo spaces should be free from defects. Unless electrical components are safe for use in an explosive atmosphere they should be isolated.

3. The ship should carry instruments for measuring:
    (a) methane in the atmosphere;
    (b) concentration of oxygen in the atmosphere;
    (c) concentration of carbon dioxide in the atmosphere;
    (d) pH value of cargo hold bilge water;
    (e) remote temperature gauge for the cargo spaces.

4. The coal should be trimmed reasonably level so as not to allow air pockets to develop.

5. The atmosphere in the cargo spaces should be regularly monitored for gasses.

6. The bilge water should be tested for pH at regular intervals.

7. Should the behaviour of the cargo differ from the declaration, this fact should be reported.

There are additional precautions for cargoes that emit methane. Mainly concerned with the venting of gas and precautions regarding the possible entry of gas into the working spaces of the ship.

The safety requirements for self heating cargoes are also spelt out. The important requirement is that the hatch covers should be kept closed, sealed if possible and only the minimum surface ventilation allowed.

Safety precautions necessary when carrying coal cargoes

All hatches filled with coal should have ventilators to allow any methane to escape into the atmosphere as it is lighter than air.

There is, however, the danger of methane from the hold finding its way into other spaces and so any enclosed space close to a hold containing a cargo of coal must be treated with care and any electrical circuits disconnected. No hazardous work such as welding should be carried out adjacent to a hatch loaded with coal. "No Smoking" notices should be displayed. Gas detection devices should be carried on board to test for the presence of methane.

The coal should be trimmed level and any side battens from the hold removed to prevent pockets of gas from being formed and also prevent air circulation through the cargo, which may cause spontaneous heating and possibly combustion.

In order to prevent methane from being produced when loading, the first few tonnes of coal in each hold must be loaded slowly, rather than being dropped from a height, this prevents the lumps from being broken up and giving off methane. It is also important to preserve the lump size as it is part of the coal's specification.

When carrying coals that are liable to heating, the temperature should be monitored by means of temperature detectors placed in the holds— any temperature above 55°C should be treated as dangerous and advice obtained. The possibility of the ship diverting to a "port of refuge" should be considered.

When carrying pond coal—ie coal that has been reclaimed after being dumped into ponds—it will obviously have a high moisture content and as it is also possible for it to have a high sulphur content it can obviously cause damage to the ship's holds from sulphuric acid.

Where there is a possibility of liquid movement the moisture content of the coal should be noted. For this type of cargo the "code" defines a transportable moisture limit as the maximum amount of moisture that the cargo can contain before it is considered dangerous. As long as the moisture content of the cargo is below this limit it can be carried safely, but should it be above this limit then special precautions as laid down in the code must be carried out.

Many coal cargoes shipped are blended (ie they are a mixture of different coals). When the blending is properly carried out this should not present any problems, but if, however the coals are loaded in layers in the ship's holds there is evidence that this can increase the danger of heating and fires as the fires appear to originate at the boundaries of the layers. Pond coal with its extra moisture content can contribute to these problems.

When coal is loaded in a wet condition, there will be an apparent loss of weight during the voyage. It is therefore important that the moisture content on loading is noted in order to counter any claims for loss in transit.

STOWAGE FACTORS FOR COAL

The stowage factors will depend on how well the coal is trimmed. In a bulk carrier this should not prove to be a problem but with a tweendecker it will be higher than for the figures quoted because of the difficulty of trimming the coal into all the spaces, which, together with the problem of size is the reason why this type of ship is not used as much in the coal trades as the larger bulk carriers.

Table 80 Stowage factors

Origin	M3/Tonne
American	1.17/1.28
New South Wales	1.08/1.25
Queensland	1.08/1.20
South Africa	1.19

It is important therefore when selecting a ship for a cargo of coal that these stowage factors are taken into account in order to ensure that when the holds are full she is carrying a reasonable deadweight.

MEASUREMENT OF COAL CARGOES

There are three main methods of measuring the weight of a coal cargo loaded into a ship. The coal can be weighed ashore by using hoppers or conveyor belt weighing before it is loaded or after it has been discharged, or it can be weighed indirectly by means of a draft survey.

The two methods of obtaining the weight of a cargo ashore are:

Conveyor weighing

In this case the coal is weighed on a conveyor belt as it is carried to or from the ship. Transducers are fitted to the conveyor belt, which sense the weight of coal as it passes over the rollers. When the speed of the conveyor belt is included in the calculation the total weight of coal passing along the conveyor can be ascertained. The chief advantage of this method is that of speed. There is no interruption of the cargo flow while the weighing takes place, the accuracy is dependent on how well the equipment has been calibrated. Improvements in computer and sensor technology are leading to increased confidence in this method.

Hopper weighing

This is considered to be the most accurate method of weighing bulks. It is mostly used for the more valuable bulks, such as grain, rather than coal. With this method three hoppers are used. The material is fed continuously into the first hopper, which then feeds it into a second hopper connected to an exact weighing device. When the second hopper is full the weight is noted and the doors will open allowing the cargo to drop into the third hopper. By using the first and third hoppers as buffers it is possible to batch weigh the cargo without stopping the flow.

The accuracy of both these methods is only as good as the calibration of the equipment—no matter how good the original calibration is, the instruments will "drift" over time. Thus cargo shortages may be caused by the defective calibration of the weighing equipment at one or both ends of the voyage.

Draft surveys

This is a method that should produce consistent results because the unit of measurement is the ship itself. The principle on which it is based is extremely simple but its success is based on a number of factors, which the surveyor may not be able to control. Perhaps the most important factor is the ability to read the draft to the required accuracy and when considering that the TPI of a 100,000 ton deadweight bulk carrier can be 225 tons, then an error of 1 inch in reading the drafts will put the calculations out by this amount.

The draft survey is based on the principle that if it is possible to ascertain the weight of the empty ship, and then ascertain the weight of the loaded ship, the difference in weights will be the amount of cargo, fuel and stores loaded into the ship. The steps for carrying out the draft survey are outlined in the following section. This is not intended to be a complete manual on the subject and readers should consult the texts for further details.

Reading the Drafts

It is important when the drafts are read that all cargo or ballast operations are suspended. This may cause difficulties with the terminal, but unless this is done, the results will not be reliable. The ship should be upright with enough ballast on board to give a positive forward draft. If either of these condition are not met there will be errors because the ship's stability information is calculated for these conditions. In particular, with the bow out of the water, which is not uncommon before the ship starts to load, the draft survey will be hopelessly inaccurate.

When reading the drafts, the three drafts on each side of the ship are read where possible, ie forward, aft and amidships. If possible a boat should be used in order to get close to the draft marks. In the case of the amidships draft marks it may be more accurate to measure the freeboard and convert it to draft. Reading drafts can be difficult if there is a swell, but there are various devices on the market that will smooth out the wave effect thus giving a more accurate reading.

Measuring the density

As soon as possible after the drafts are read the density of the water should be ascertained. It is important that there is no delay because if the port is in a river or estuary, the density will change according to the state of tide. A sample container with a perforated lid should be used and it is lowered at a steady rate to a depth equal to that of the ship's draft and then brought to the surface again. As long as the container is not full when it reaches the surface a representative sample of the dock water will have been obtained. A number of samples from the offshore side of the ship should be taken and the results averaged. Avoid taking samples from the onshore side as it is possible that water of a different density has been trapped and this will give a misleading result.

The only accurate density meters are made of glass, avoid using a brass instrument. Most measuring instruments are calibrated for water in a vacuum and 0.0011 should be subtracted from the readings to correct them for the buoyancy of the air. It is a common mistake to correct the density of the water for temperature; the reading on the instrument is the density in which the ship is floating and this is the density that must be used in the calculations.

Correction to the perpendiculars

The ship's stability information is calculated for a draft measured at the ship's perpendiculars. Thus the drafts read during the survey will have to be corrected. The forward perpendicular is a line at right angles to the keel cutting the summer waterline at the stem. The after perpendicular is a line at right angles to the keel passing through the rudder post. In most cases the draft marks painted on the ship's side do not coincide with the perpendiculars so the draft as read at the draft marks has to be corrected.

The drafts then have to be corrected for hog and sag. There are a number of methods for calculating this correction (the main methods are outlined in the references), but in general the mean of means correction is only approximate. If the surveyor has access to a computer and it is possible to read the draft at five instead of three points on each side of the ship, the accuracy can be considerably improved. From these corrections, the mean draft is obtained and the corresponding displacement can be read from the ship's displacement scale.

The corrections for trim and heel must then be applied to the displacement. There is a further source of error since the calculation of the corrections for heel and trim are approximate only, and in any case the trim should not exceed 1% of the ship's length. Some ships are provided with correction tables and if these are available, the surveyor can have more confidence in the results. Finally the displacement is corrected for the density of the dock water.

There are two methods of carrying out the draft survey. The before and after method is such that the displacement of the empty ship is obtained prior to loading, the displacement is again obtained after loading and, after taking account of any fuel and stores taken during loading, the cargo loaded can be calculated.

This method is potentially the most accurate, but there is one problem. In order to provide the ship with an acceptable trim for the first survey it is necessary to leave some water in the ballast tanks. The amount of water has to be calculated and its density measured. Because of errors in the calibration of these tanks, errors can occur when measuring the water content in a partially full tank. To avoid this, where possible one or two tanks should be left full rather than a number of tanks partially full.

The second method is to measure the displacement of the loaded ship and then deduct the weights of the fuel, stores, water and constant in to arrive at the weight of the cargo. This method is potentially less accurate but it may be necessary because either the surveyor was not in attendance when the ship first arrived or because the terminal is not prepared to accept the delay to the cargo operations when the draft survey is being carried out. The problems with this are in obtaining the exact weights of the fuel, water and other items on the ship. The shipowner is paid freight on the amount of cargo loaded so it will always be in his interest to maximise the weight of cargo loaded. Thus the weight of fuel and water must be measured, do not just take the ship's figures for granted. The "constant" (see page 120) is the difference between the calculated deadweight corresponding to the draft and the actual deadweight. In practice, although a contradiction in terms, the constant can vary as it represents all the items on board the ship that have not been accounted for.

Sources of error in draft surveys

When properly carried out the draft survey can be as accurate as any other method of measuring the weight of a cargo of coal, but there are a number of problems, some of which are outside the control of the surveyor, and these can degrade the accuracy of the survey.

1. *Mud in the ballast tanks*: Water ballast is frequently loaded in a harbour or river. In these circumstances the water will contain mud in suspension. During the voyage the mud will settle to the bottom of the tanks and over a period of years a considerable deposit will build up. This will lead to an increase in the weight of the "constant", which can affect the accuracy when the surveyor only carries out the survey on the

loaded ship. This is the reason why surveys should be carried out on the loaded and empty ship.

2. *Squat*: When a ship is in a fast flowing river, the draft increases due to the action of the "bernoulli" effect causing a reduction in pressure under the hull. In situations where the underkeel clearance is small and the current exceeds two or three knots, the draft survey can seriously exaggerate the weight of cargo lifted.

3. *Errors in reading the Drafts*: The accuracy of a draft survey is crucially dependent on how accurate the drafts are read. In some terminals it may not be possible to obtain a boat and if the sea is disturbed, then the accuracy will suffer. Special equipment is available to smooth out the effect of waves on the drafts.

4. *Errors in the ship's calibration tables*: The ship's hydrostatic particulars are calculated from the plans, not the actual ship. In many shipbuilding yards it is not practice to measure the new ship, so the accuracy of the tank calibrations and other data depends on how closely the plans were followed.

These and other possible sources of error mean that the surveyor's report should be comprehensive—a bare statement of the cargo quantity on board is useless. The report should include the following:

(a) Which methods were used for the survey.
(b) Whether a boat was used for reading the drafts.
(c) Sea conditions.
(d) Which values were estimated.
(e) Current and depth of water when the survey was undertaken.

The papers on the Uniform Code of Standards (ref) give a useful pro-forma which can be used for this purpose.

With this type of information it is possible to make a reasoned judgement as to the accuracy of the draft survey.

Despite these difficulties and other causes of error, draft surveys are used for a large number of coal cargoes as they have the important factor of consistency. Assuming that the cargo is measured by draft survey in both the loading and discharge ports, there should be close agreement between the two figures. A ship that is surveyed frequently should have up-to-date values for the constant and a record of draft surveys should be maintained on board for the use of surveyors.

It is surprising, considering the importance of draft surveys and the need for co-operation between the surveyor, the ship's officers and the terminal authorities, that there is no clause in any of the standard charter parties requiring that the ship's officers undertake the following:

(a) Ensure that the ship arrives with a suitable trim for draft surveys.

(b) Requires the master to furnish the surveyor with the required particulars.
(c) Requires the master to assist the surveyor when carrying out the survey.
(d) States whether time used for the taking of draft surveys is to count as laytime, and how time spent in preparing the ship for draft surveys is to be treated.

The papers on the performance of draft surveys provide a possible charter party clause to cover these points.

REFERENCES FOR CHAPTER 7

Thomas' *Stowage*
International Maritime Organization, *International Maritime Dangerous Goods (IMDG)*.
"The Uniform Code of Standards and Procedures for the Performance of Draft Surveys", Conference held in Katowice, Poland, May 1990.
Nautical Institute, *The Work of the Nautical Surveyor* (London).
International Maritime Organization, *The Code of Safe Practice for Solid Bulk Cargoes*.
W.F. Berry and J.S. Goscinski, *Heating of coals in Transit—Problems, Causes and Solutions* (Seaways Jan 1983.)

CHAPTER 8

Chartering

In the coal trade, the method by which traders arrange for the cargoes to be moved is by chartering sufficient ships for their requirements. The nature of the trade means that shippers need to be able to obtain some ships to meet their long-term requirements while also being able to obtain ships for single voyages to cover the peaks and troughs. The traditional method for filing these requirements is by chartering. There are a number of different types.

TYPES OF CHARTER

There are a number of different methods of chartering a ship, which are used by the traders to ensure that their long- and short-term requirements can be met in the most economical manner.

1. Voyage chartering

With a voyage charter party, the ship is employed on a single voyage, ie from a certain port to a port or ports in an agreed area.

The shipowner will be paid freight, either per tonne of cargo or as a lump sum, which is normally payable either at the destination, or on signing the bills of lading.

The amount of cargo to be loaded is agreed in advance. The usual method is for the charterer to provide a full cargo, but as the owner does not know at this stage exactly how much cargo the ship can lift it is usually described as a given tonnage with a fixed percentage (5%) more or less owner's option (MOLOO). For example, 50,000 tonnes 5% MOLOO means that the owner can lift 50,000 tonnes + 2,500 (5%) = 52,500 or 50,000 − 2,500 = 47,500. On arrival at the loading port the master will declare how much cargo he is able to lift. If the charterer cannot provide this quantity he will be expected to pay deadfreight— ie to compensate the owner for the loss of earnings resulting from not being able to provide the full cargo.

When the charterer is not certain whether there will be enough cargo to fill the ship, the freight may be paid as a lump sum. This means that the charterer can load as much or as little cargo as he wants, there is no contractual obligation to provide a full cargo and so the question of deadfreight will not arise.

The amount of time for loading and discharging the cargo is agreed; this is known as laytime. Should the ship be delayed in port due to lack of cargo or other causes that could reasonably be said to be the charterer's responsibility then the shipowner is entitled to claim compensation in the form of "demurrage", while if the ship should finish earlier than expected, the shipowner pays dispatch to the charterer.

Dates are then fixed. The charter party will state, for example, laycan 25/30 April, the abbreviation "laycan" meaning laydays, cancelling date 25/30 April. This means that the ship must arrive on or after 25 April and before 30 April. The charterer can refuse to load if the ship arrives before the first date. In fact the cargo may not even be available. On the other hand, if she should arrive after the cancelling date the charterer is at liberty to cancel the ship and find another. This would prove to be a useful option should freight rates have fallen since the charter was fixed. However the converse is not necessarily the case—if a ship is late arriving and rates have risen in the interim, should the charterer indicate that the ship is still wanted, then the owner must proceed with the voyage.

The shipowner is responsible for all the expenses of running the ship and also the additional voyage expenses apart from the cargo handling costs, which in the coal trade are usually paid for by the charterer. See, however, references to clauses in the charter parties regarding overtime costs when loading or discharging the cargo.

The owner calculates the cost of the proposed voyage and compares this with the anticipated freight to decide whether the voyage will be profitable. It is important that the owner knows how long the intended voyage will take as the calculations are based on profit per day. Hence a delay may make the difference between a profitable or a loss-making voyage. If the delays beyond the allowed laytime are the fault of the charterer then demurrage will be paid as liquidated damages to the owner; but should the ship load or discharge faster than anticipated then dispatch will be paid by the owner to the charterer.

The owner always remains the carrier and when the master signs the bills of lading they are signed on behalf of the owner. This means that in the event of any claims for shortages or other discrepancies in the cargo, the owner is responsible and not the charterer.

2. Consecutive voyages

It is possible for a charterer to fix a ship for a series of round voyages. In such a case each voyage is considered a separate entity as far as freight and demurrage are concerned. Clauses are available to protect the owner should fuel prices or other costs, for example war risks, change during the contract.

3. Part charters

It is possible where one charterer is not able to provide a complete cargo, for the owner to arrange a number of charters, and there will be a separate charter party for each of the different parcels that are carried. It is important therefore that these charter parties are drawn up so that there is no conflict between the various interests involved. Demurrage can be a problem with this type of charter.

4. Time charters

This is where the charterer takes on the ship for a certain period of time. It could be as short as one month or as long as 20 years, although the latter is not very common. The shipowner still operates the ship but instead of earning freight, he is paid hire at an agreed amount either per day or per month, either as a lump sum or as so much per deadweight ton per month (eg New York Produce Exchange Form).

> 4. The charterers shall pay for the use and hire of the said vessel at the rate of   daily or   United States Currency per ton on vessel's total deadweight carrying capacity, including bunkers and stores on   summer freeboard, per calender month commencing on and from the day of her delivery, as aforesaid, and at and after the same rate for any part of a month; hire shall continue until the hour of the day of her redelivery in like good order and condition, ordinary wear and tear excepted, to the Owners (unless vessel is lost) at   unless otherwise mutually agreed. Charterers shall give owners not less than   days notice of vessels expected date of redelivery and probable port.

The charterers are responsible for finding the cargo and employing the ship including port, canal, cargo handling, hold cleaning and fuel costs. The management of the ship, however, remains the responsibility of the owner.

A charterer may sublet the ship, ie operate her on the voyage or other markets. In this case he becomes the disponent owner, which means that although he does not own the ship, he is entitled to the benefits that are obtained from trading her during the period of the charter. The charterer is the carrier and the master signs the bills of lading on his behalf. Regardless of the number of sub charters, the head or first charterer is always responsible to the owner for the employment of the ship.

Examples of time charter forms are the New York Produce Exchange form and the BALTIME. The former is produced by the Association of Shipbrokers and Agents, New York (ASBA), and considered to favour the charterer while the latter is produced by BIMCO and is considered to favour the owners.

5. Time charter trips

These are a combination of the voyage and time charter and may best be described as a voyage charter on time charter terms. Although the charter is usually for a single voyage, the division of responsibilities is the same as those for a time charter and the normal time charter forms are used. These charters may be used by shipowners where port delays are expected or during periods of uncertainty over fuel prices and availability, where, because of his greater market presence, the charterer is able to secure supplies of fuel.

The advantage of a trip charter as far as the charterer is concerned is the greater freedom that this allows with regard to the details of the voyage. The charterer is able to select ports or indeed a trading area without having to agree the details with the owner. This is because the latter is indifferent to any additional costs that may result from the charterer's choice of ports, as these costs will be met by the charterer.

6. Bareboat (demise) charters

With this type of charter, the charterer both manages and operates the ship; the shipowner (possibly a financial institution) virtually giving up control for a fixed period of time. The charterer is the disponent owner responsible for both crewing and managing as well as employing the ship.

The owner will receive the fixed rate of hire at regular intervals during the period of the charter.

Bareboat or demise chartering is frequently used in ship finance where it may suit a bank to own the ship, frequently because of tax advantages, but they don't want to get involved in running it as they do not have the necessary knowledge or experience. At the same time the disponent owner can manage the ship profitably but the current market rates will not repay the full capital costs of the ship within the period of a bank loan. There is frequently an option to purchase at the end of the demise charter. Two charter parties used for this type of charter are BARECON "A" and "B", bareboat charters produced by BIMCO. There is also a version of BARECON B for ships building in Japan.

7. Contracts of affreightment (COA)

These are used where a shipowner or operator agrees to transport a given quantity over a fixed period of time. The main difference between this and conventional chartering is that the choice of ships is left up to the shipowner. Thus although it is possible that a list of proposed ships will have been presented for the charterer's approval, the owner is free to select whichever ship is free for a particular voyage. This type of contract gives the owner considerable freedom to manage his fleet to the best advantage, even to the extent of chartering ships in if the owner's fleet is engaged in more profitable employment elsewhere. There are few documents used for this type of contract as all that the charterer requires is an assurance that the quantities required will be shifted using approved ships. The rest of the details are up to the owner. This type of contract is common with small coasters as it saves having to fix a ship for each movement.

The standard documents for this purpose include Volcoa, which is basically for dry cargo, and Intercoa, which was written for oil cargoes but can be adapted for dry cargo with the alteration of a few lines. These forms are designed to be used in conjunction with voyage charter forms for each voyage that is undertaken under the COA.

THE DECISION WHETHER TO TIME OR VOYAGE CHARTER A SHIP

There are a number of reasons that would affect the selection of a particular type of charter by either an owner or charterer.

The voyage charter is favoured by the trader with the "one off" cargo or by the utility or steel company who need to supplement a fleet of owned or time chartered ships. In the coal trade voyage chartering is used by traders who ship "speculative" cargoes. These are cargoes where the final destination has not been fixed. The trader loads a cargo for a port or range of ports and when the cargo has been loaded and the ship has left the loading port, then sells the cargo. A cargo may change hands several times during a voyage. It is important therefore that all the terms agreed during the chartering fixture are clearly spelt out so that the eventual buyers of the cargo (who were not involved in the original negotiations) will understand what their responsibilities are under the charter party.

The shipowner will select a voyage charter for the potentially higher profits that can normally be obtained on the "spot" market or when he wants to keep a ship available for a possible sale.

The time charter is used by some traders as a means of controlling costs. With a fleet of ships on time charter, the trader is protected from the freight rate fluctuations that are a feature of "spot" market trading.

The shipowner may also prefer time charters for the greater security that comes from knowing that the ship has secure and profitable employment for a known period of time, which may be as long as several years.

The time charter is frequently used as part of the security for bank loans when financing a ship. The second hand price of a ship varies according to the freight rates and so the ship itself is not considered sufficient security for the loan. The bankers like the secure earnings of a time charter over at least some of the period of the loan. It is possible to design ship financing packages to bridge the gap between current freight rates and the high capital costs of new ships using time charters.

The decision on how to charter a ship is also based on how the market views the current freight rates. When rates are low and likely to rise, owners will be looking for short-term contracts so as to leave their ships free to seek more profitable employment when the rates have risen. In this situation owners will therefore not accept time charters unless they are above the current market rate, something that the shippers are not likely to agree to. Conversely, when rates are high and the perception is that they are likely to fall, owners will seek time charterers to retain the higher earnings for as long as possible, while shippers want to keep their contracts as short as possible. Thus time charters are only seen when the market is in a relatively stable position and both parties are willing to enter long-term contracts.

CHARTER PARTY FORMS

The long history of chartering for coal cargoes means that there are a large number of charter party forms in use. Many of these, such as the Baltic and International Maritime Conference (BIMCO) POLCOALVOY are written for a specific trade, and therefore deal with the customs of the ports involved, while others are suitable, with additional clauses, for any voyage.

There is a strong tradition in shipping to stick to tried and trusted forms, so many of the charter party forms are old and do not properly take the massive changes brought about by the use of large bulk carriers into account. They therefore have to be modified by the inclusion of additional or side clauses. This leads to complications and disputes over the meaning of these clauses, particularly where they have been badly drafted, which may be the case when they have been made up on the spur of the moment. To alleviate this, organisations like BIMCO have drawn up a number of charter party forms to avoid the use of additional clauses.

The role of BIMCO in chartering

The Baltic and International Maritime Conference was founded in 1905 under the name of the Baltic and White Sea Conference by a group of shipowners interested in the coal and timber trades to the Baltic and White Seas. It is now international in its interests and its objectives are:

1. To unite shipowners and other interested groups (brokers P&I clubs) connected with the shipping industry.

2. Communicate to members about unfair charges or objectionable practices.

3. Prepare and improve charter parties and other shipping documents and to do so whenever possible by means of friendly negotiation with charterers, shippers, merchants receivers, shipowners, shipbrokers and others connected with the shipping industry or with organisations representing any such persons, and to exhaust any reasonable means of agreement before issuing any forms for use by the shipping industry.

4. To issue as approved documents for the use of the shipping industry, charter parties and other shipping documents, and to adopt as approved documents for the like use, charter parties and other shipping documents that have been issued by similar organisations or agreed by representatives of the parties concerned.

5. To meet and/or correspond with charterers, shippers, merchants, receivers, shipowners, shipbrokers and others engaged in the shipping industry, and with representatives or organisations of any such persons, as to any matter connected with the shipping industry.

6. Take such steps in the interests of the shipping industry as appear desirable.

BIMCO keep their members informed of the latest developments by publishing a monthly journal and various books on chartering. Their book *Check Before Fixing* is a useful document for anyone proposing to charter ships, and their monthly journal is useful for keeping members up to date with the latest court decisions.

Types of charter party form

An "official" charter party form is a form that has been agreed and passed by an official body such as BIMCO. These are distinguished from other forms that are not considered to be as suitable for use, possibly because they unduly favour one or other of the parties involved. The shipowner is advised whenever possible to use these forms because they are usually well drafted and balanced.

The "official" forms may be described as:

1. *An agreed charter party*: this is a form that has been agreed between BIMCO or another group of shipowners and charterers for the trade concerned. The terms must not be altered in any way. The use of this form is compulsory for that particular trade. Examples of agreed forms and Polcoalvoy and Sovcoalvoy.

2. *An adopted charter party*: if a charter party has been agreed by two organisations, and another organisation wishes to make use of it they may then adopt the charter party, thus BIMCO may adopt a charter party agreed by the Chamber of Shipping of the U.K. and a number of charterers. BIMCO members would then be expected to use this document wherever possible.

3. *A recommended charter party* is used where no specific charter party exists for a particular trade and an organisation like BIMCO would recommend a particular charter party, such as Gencon, for use by its members in that particular trade.

Some of the charter parties and other forms used in the coal trades are listed below.

Voyage charter party forms

A number of general purpose charter party forms are available that could be used for coal cargoes. Some common forms are:

- *Gencon**: this is a general purpose voyage charter party, which is considered fair to both parties.
- *Universal Voyage Charter party 1984 (NUVOY–84) Multiform**: this is a general purpose charter party issued by FONASBA (Federation of National Shipbrokers and Agents).

The specialised coal charter party forms in common use include:

- Polcoalvoy*
- Sovcoal*

There are also Polcoalbill* and Sovcoalbill* bill of lading forms to go with the respective charter parties.

Americanised Welsh Coal Charter (Amwelsh)
Australian Coal Charter
Safanchart No 2
Safanchart No 1

Time charter forms

There are no specific charter party forms for coal cargoes, but two common forms are:

- New York Produce Exchange
- Baltime*

Bareboat charter forms (demise)

- Barecon A*
- Barecon B*

Contract of affreightment forms

- Intercoa
- Volcoa*

* Indicates that forms are issued or recommended by BIMCO

The balance between owners' and charterers' interests in a charter party form

For many years the clauses in a bill of lading have been subject to statute law. The defences that a ship owner may claim in order to avoid paying compensation for damage to the cargo that is entrusted to his care for the voyage are limited by international conventions. The latest of these conventions have been codified into the Hague–Visby rules. These are incorporated into national legislation as the Carriage of Goods by Sea Acts. The purpose of these Acts is to protect the shipper who may have little experience of the perils of the sea or indeed of shipowners, from a shipowner who could, and certainly did make use of the monopoly powers that the conference system provided to avoid any liability for damage to the cargo.

The construction of charter party forms on the other hand, has always been left up to the owners and charterers to agree, the assumption being that both shipowners and charterers understood the complexities of chartering and could therefore be left to draw up a charter party without the need of any statutes to provide protection. The consequences of this are that when a charter is being negotiated, the choice of charter party form is one of the first items discussed in the negotiations. When freight rates are high, the owner is able to obtain what he considers to be a fair charter party, or can at least get the most objectionable clauses removed from the charter's preferred form. When the market favours the charterer, the charterer will be able to impose his favoured form without any modification.

This situation may change in the future. Talks are now in progress at UNCTAD between a number of interested groups on the subject of statutes governing the content of charter parties. Should these talks bear fruit then a number of biased charter party forms will be heavily modified or vanish altogether. Thus the BIMCO approved forms are generally considered to be fair to both parties while some of the other forms are considered to favour the charterer to a greater or lesser extent.

Additional clauses

With a general purpose charter party, it is most unlikely that a voyage be agreed without any additional clauses (known as side clauses) being added. These clauses always take preference over the printed ones in the charter party. Although these clauses may be drafted by the owner or the broker to cover a particular situation, there are a number of recognised clauses available. The advantage of using such clauses is that they are well known and their meaning is usually understood by both parties. This presents less risk of a dispute following a misunderstanding of a particular clause than would be the case if it had been specially drafted for the occasion.

These clauses are used where additional costs that are not mentioned in the charter party document are incurred. An example is in American ports where the owner will want to include "Dumping and trimming" costs to be paid by the charterer. With older ships the cargo had to be trimmed ie levelled off, before the ship could proceed to sea. Although it is not as important now, these costs are frequently included in the port costs and it needs to be spelt out who is to pay for them.

One of the difficulties with chartering ships to carry coal is that the common forms like the Amwelsh and the NYPE form are out of date and before a charter can be fixed using these forms a large number of additional clauses must be agreed, which makes fixing on these forms a time consuming business. The problems with these forms are discussed in more detail in the following sections.

CHARTER PARTY FORMS USED IN THE COAL TRADE

As the reader would expect, with such a long tradition of carrying coal by sea, a large number of voyage charter parties are in use. The charter parties that are to be discussed here are:

1. Americanized Welsh Coal Charter 1953.
2. Australian Coal Charter.
3. Baltic and International Maritime Conference (POLCOALVOY).
4. Safanchart No 2.
5. Safanchart No 1.

It is proposed to compare the different charter parties under a number of headings.

Commencement of laytime

The commencement of laytime in these charter parties dates from before the advent of bulk carriers and modern cargo handling equipment when coal was loaded and discharged by bucket and shovel. Today with modern bulk terminals, should the ship be able to berth on arrival, it is usual for a ship to have finished loading before the laytime starts, while in discharge ports a discharging rate of 40,000 tonnes per day is possible and rates of 25,000 tonnes per day (depending on ship size and cargo quantity) are common. It is therefore important that when fixing ships using these charters, the probability of having to pay dispatch is taken into account. Thus, depending on market conditions the owner should either refuse to accept dispatch or, if that is not possible because of market conditions, then at least allow for it in the voyage estimates.

All the charter parties have a waiting period between submitting notice of readiness and when laytime starts to count. This is a relic of the old days when communications were not as efficient as now and it was necessary to organise the labour after the ship's arrival before starting to load or discharge the cargo. With coal being handled at high speeds in modern loading terminals, cargo operations now start as soon as the ship is berthed, so this time is no longer necessary. In many cases this has an important effect on the laytime allowed, as it is possible in a modern terminal for a ship to be loaded within the notice of readiness period. The result of this is that the owner could end up paying a considerable amount of dispatch unless these clauses in the charter party are amended by side clauses.

"Turn time" is another method used by the charterer to increase any dispatch payable. It is used with the Amwelsh charter party and frequently adds an additional 12 hours after the notice of readiness time has expired before laytime can start at both loading and discharging ports.

1. Americanized Welsh Coal Charter

Laytime will start to count 24 hours after notice of readiness to load has been presented, between 9 am and 5 pm, Monday to Friday and 9am and 12 noon on Saturday. The clause then goes on to state that unless sooner berthed time will start on receipt of notice. With the discharge port, time will start to count 24 hours, (Sundays and holidays excepted) after notice has been presented. There is no mention of notice of readiness being presented during office hours.

2. Australian Coal Charter

Laytime will count 24 hours after notice of readiness has been presented between 9 am and 5 pm, Monday to Friday and 9 am to 12 noon on Saturday. The charter goes on to specify that the same conditions apply to the discharge port.

3. Polcoalvoy

Laytime starts at midnight following notice of readiness if tendered before noon, and 7 am the next day if notice of readiness is tendered in the afternoon. Notice is to be tendered in ordinary office hours (ie between 8am and 4 pm on a working day).

4. Safanchart No 2

Laytime will start to count 12 hours after notice of readiness has been presented between 7 am and 4 pm, excluding Saturdays, Sundays and holidays. In the discharge port the same 12 hours' notice time applies but the charter states that notice must be tendered in accepted business hours, no time being specified, and there is no mention of holidays or weekends.

5. Safanchart No 1

Notice of readiness accompanied by a surveyor's certificate is to be handed in business hours in between 7 am and 4 pm (excluding Saturdays) and laytime will start to count 24 hours after notice of readiness is received. With the discharge port, time starts to count 24 hours after notice of readiness to discharge has been given in accepted office hours.

Thus there is considerable difference between the charters as to when time starts to count, with the Safanchart being the most generous to the shipowner. However, when comparing the different charter parties it is perhaps more useful to consider what is the owner's position should the ship start cargo within this period.

Laytime

All the charter parties count the laytime in running days of 24 hours. Thus one day will not be completed until 24 hours have been spent loading or discharging. This allows for the fact that many loading terminals work around the clock and so the time worked will actually count.

All the charter parties exclude Sundays and Holidays (SHEX). This may be changed and a number of coal charters are fixed SHINC (Sundays and holidays included). The assumption is that with a modern terminal these times would be worked if the ship was alongside. The treatment of other excepted times vary.

Thus the Amwelsh charter party allows time to count as soon as the ship starts to work and there is no penalty on the owner for starting to load as soon as the ship is ready. Sundays and holidays are excluded and so is time after noon on Saturday and the day previous to a holiday, and up to 7am on the day after a holiday but the charter party allows for any time used to count as laytime, although it goes on to state that only time actually used to load the cargo is to count. Thus with this charter there is no penalty on the owner if he uses what would normally be excepted time to load or discharge cargo. The discharging clause is similar to the loading except for the addition of the words weather permitting, although this is also one of the exceptions in the loading clause.

The Australian coal charter only allows one third of the time worked prior to notice of readiness or in notice of readiness time to count as laytime, provided it is not in one of the excepted periods. With this charter any time worked on Saturday after 12 noon and Sundays or holidays is not to count although time starts to count again on these occasions immediately after midnight. Thus this charter party could land an owner with a considerable expense for dispatch if the ship were to load a cargo over a holiday period.

The Polcoalvoy charter party allows any time worked during the notice of readiness period to be counted in full, and also recognises the more modern methods of carrying coal by allowing for SHINC terms (Sundays and holidays included) to be used. The remainder of the charter parties would have to be amended for this. Because it is designed for Polish coal, the holiday periods are spelt out—ie Sundays, legal holidays and 4 December are given.

Exceptions

There are a number of delays that will not count as laytime, the obvious one is that of equipment breakdown when the ship is using her own gear to load or discharge cargo. The shipowner can hardly expect sympathy when the breakdown is due to his inability to maintain the ship's equipment. There are, however, exceptions that are less fair to the owner.

Strike clauses

These are to cover strikes by stevedores and contractors, not the ship's crew. The important point is to define what a strike is. It is possible for a "work to rule" or other industrial action short of a strike to delay the ship without the shipowner being able to gain any protection provided by these clauses. The Amwelsh form does not allow time lost resulting from a strike to count unless the ship is already on demurrage. The Polcoalvoy covers this in more detail and is much fairer to the owner. In the event of a strike at the loading port, provided the ship has not already arrived, the owner can ask the charterer to count as laytime any delays that might arise as a result of the strike. If the charterer is not willing to pay for these delays the owner has the option to cancel the charter. Should the ship be partly loaded before the strike, the ship must carry the cargo already loaded to the discharge port, but the owner has the liberty to load any further cargo to reduce the loss.

A more difficult situation occurs when the ship is in the discharge port with cargo on board. In this situation the consignee has the option of holding the ship until the end of the strike, and paying for any delays that are incurred as a result of the strike, at half the demurrage rate. The consignee has the option of ordering the ship to another port for discharge. In this case no extra freight is incurred unless the distance to the new port exceeds 100 miles. Thus in this charter the responsibilities and liabilities of both parties are clearly spelt out and the costs of the strike are shared between the owner and charterer. The other charter parties do not cover the situation in any detail, some merely stating that time lost is not to count but there is no mention of alternative discharge ports. Thus in the event of a prolonged strike it would be up to the owner and charterer to come to some mutually agreeable arrangement.

Ice

There are two difficulties where ice is concerned. The first is that it may be difficult or impossible to reach a port because the port or approaches are obstructed by ice. This could be extended to where port is clear but the navigation marks have been removed because of the ice. The other situation is where a ship is already in port and the onset of ice threatens to prevent the ship from leaving. This situation is made more difficult because navigation in ice requires special skills, which the average ship's master would not be expected to possess. There is, however, one important difference between ice and other problems that may cause difficulties during a charter, and that is if an owner fixes a ship for a port or ports where there is risk of ice, he is, or should be aware of the risk and is therefore not in a position to blame the charterer should things go wrong.

The problem of ice is dealt with by the different forms in the following way. The Polcoalvoy, by nature of its trade, covers this problem in considerable detail. Should the master consider that there is a danger of ice in the loading port or ports, then he is always at liberty to sail. This would make the charter void unless there is cargo already on board, in which case he must proceed to the discharge port.

At the discharge port, the master is again free to refuse to enter the port but must ask the charterer to nominate an alternative discharge port, no extra freight being payable unless the distance between the two ports exceed 100 miles.

The Australian Coal Charter has a similar discharge clause, but the other forms do not mention ice, so it is up to the owner to ensure that a suitable clause is incorporated if the ship is to proceed to a port where there is the risk of ice being present.

Other causes of delay

The various charter party forms are clear about what delays will count towards laytime. The Polcoalvoy exempts the charterer from the consequences of any delays other than those mentioned in the strike and ice clauses, although there is a break—if the vessel is held up for more than four days the owner can cancel the charter. There is also a clause to deal with the position where there is a part cargo on board. The other forms are not as clear. The Amwelsh form treats all delays as the same and after six corrective days delay, the charter is null and void provided no cargo has been loaded. The other forms merely state that for delays that are not the responsibility of the charterer, laytime is not to count.

Overtime

When a modern terminal is used it is unlikely that the owner would have the option of refusing to work cargo working during an excepted period. This is because the terminals would quickly become congested. Therefore the terminal operator or port authority will decide when the ship will work cargo. This is taken into account in the various charter parties in the following manner:

1. *Ships officers and crew*: the cost of any overtime that the owner may have to pay the crew is down to the owner's account.

2. *Other costs*: all the charter parties start by stating that these costs are for the party ordering the overtime. This is fair as far as it goes because those parties who order the overtime are assumed to benefit from it. The difficulty arises when the overtime is ordered by a port or terminal authority. In this case the costs will have to be paid by the owner or the consignee. The Polcoalvoy is silent over this situation, while the other forms (with the exception of the two South African forms) accept that it is the charterer or receiver who pays. The latter two forms split the costs equally between the owner and charterer/receiver. The important point that arises from an owner's point of view is that when using these forms, he could be penalised twice. First by the cost of his share of the overtime bill should it be ordered by the terminal owner who is likely to be different from the consignee. Secondly, he could find that the result of the overtime is that the ship is now finished much sooner than anticipated and he ends up paying dispatch to the charterer.

To appreciate the problem of demurrage/dispatch and the question of overtime it is important to appreciate that a shipowner works out his income on a daily basis. Thus, all things being equal (such as having further profitable employment for the ship), the owner stands to benefit from a faster than anticipated completion of the voyage. For the charterer on the other hand, the important consideration is cost per ton. It is unlikely that an extra day or so on the voyage will make much difference unless there is the possibility of the ship missing a deadline in the contract of sale for the cargo. So provided the amount of dispatch negotiated in the charter party is reasonable and fair, both parties stand to gain from a quick turnround in port.

Several berths in one port

In some cases the charterer may want to load or discharge at different berths in the same port. This could be because the cargo is split between a number of different consignees or just because there are a number of different grades of coal on board and they have to be delivered to different locations. When fixing the charter, the charterer will want to be free to discharge at as many different ports/berths as possible as this will increase his chances of making a profit by selling all or part of the cargo.

There are two different aspects to this problem:

1. Who pays for the shifting, tugs pilots, etc?
2. Does the time spent in shifting count as laytime?

All the forms allow for a number of different berths in the same port to be used for discharge. The Polcoalvoy allows for two or more berths in the same port to be used for loading or discharging, with time spent shifting to count as laytime but the cost of shifting to be for the owner's account. The other charter forms vary. The South African forms do not allow time to count unless more than two berths are used, while in the Australian form, cost of shifting is for the owner's account but time is allowed to count as laytime. The Amwelsh form is silent over this point. It is therefore important that an owner takes this into account when undertaking the voyage estimate prior to agreeing the charter. Where the charterer agrees to pay the costs of shifting the owner should arrange for the charterer to pay these costs directly to the parties concerned.

The charter forms all give the option for loading or discharging into lighters. This would occur where a berth is not available for the ship but it can be a time consuming process and the forms explicitly state that demurrage will be paid should the allowed laytime be exceeded. A point that is not covered but is made in other parts of the charter party concerns the use of ship's gear. In general, coal is loaded and discharged using shore appliances because they are much faster than ship's gear. Where a ship is anchored out in the harbour and using lighters, floating discharge equipment is usually provided, but occasionally the ship's gear must be used. The type of cargo handling gear is mentioned under the ship's description and owner warrants that it is suitable for discharging coal. In some cases the charterer will place grabs on board the ship in the loading port, which will then taken to the discharge port and used for discharging the cargo.

Extra insurance

Depending on the terms of trade—INCO terms—cargo insurance may be paid by the shipper or the receiver. Should the ship be over 15 years old, it is possible that the underwriters may want an additional premium because of the age of the ship. Most of the charter parties have clauses that state that this is paid by the shipowner. In conditions of high freight rates it may be possible for the owner to have such a clause deleted, but it is a problem that owners and charterers should be aware of as there are a large number of bulk carriers built between 1971 and 1975 engaged in carrying coal. In such a case the additional costs must be included in the voyage estimate and the owner should ensure that the charter party is claused to the effect that the rates are as per Lloyd's of London. Otherwise the owner may not have any control over the additional premium charged.

War risk

The need for such a clause has been underlined by the recent Gulf conflict. The traditional type of clause covered the case where a port was blockaded as a result of a war but made no mention of the situation where a war has not been declared. In recent years this type of situation is more typical than where there is an actual war. Another problem that a war clause needs to address is that of the additional insurance premiums. In many cases the war risk premium is likely to change at short notice and the shipowner could end up having to pay a much higher premium than was envisaged when the charter was signed.

Stowage of the cargo

As discussed in Chapter 8 it is possible to damage a ship by incorrect stowage of cargo. The charterer is clearly not responsible for any damage that might result from incorrect stowage of the cargo but there are a number of clauses in the charter parties that clarify this point.

One of the problems that might arise is that as the charterer pays the stevedores to load or discharge the cargo, there are therefore clauses designed to limit the charterer's responsibility to that of paying the stevedores. The clause in Safanchart No 1 makes this point clear:

Stevedores and/or Trimmers. Stevedores and trimmers for loading, trimming and discharging to be employed by Charterers or Shippers/Receivers at their expense and under Master's control. Stevedores and/or Trimmers shall be considered as Owners servants and the Charterers/Shippers/Receivers are not to be responsible for any negligence, default or error in judgement of the stevedores and/or Trimmers employed in loading, and/or discharging.

Similar clauses are contained in all the charter party forms and the intention is clear. Despite the fact that the stevedores/trimmers are paid by the charterer and the owner is forced by the nature of the voyage to use them, the master is responsible for the safe loading, trimming and discharging of the cargo. In addition to such clauses the owner should insert a clause requiring the charterer to assist in the recovery of any claims, as it is he who pays for the stevedores.

Grabs

Coal is loaded and discharged by mechanical equipment, and there are two clauses that cover the problems that might arise from the use of this equipment.

The first clause relates to where the cargo is stowed. In order for a cargo to be rapidly discharged it must be possible for it to be reached by the grabs. There must also be a provision for damage caused by the use of the grabs to be ascertained and repaired.

The clause on grab discharge in the Safanchart No 1 covers most of the points:

Owners warrant that the vessel is in every way suitable for the entire cargo to be discharged by Grabs. Should owners be in breach of this warranty and should cargo be loaded and trimmed in deeptanks, tweendecks or bunker spaces or in any area not readily accessible to Grabs (hereinafter referred to as "inaccessible areas"), any and all extra expenses and any time lost in such loading and trimming and any and all extra expense and any loss of time in such loading and trimming and all extra expenses over and above the cost of normal Grab discharge and any and all time lost by reason of loading into and discharging from such inaccessible areas shall be for Owner's account. Charterer's Representatives, agents and Master to issue a joint statement at discharge ports stating:

(a) tonnage (if any) actually discharged from inaccessible areas.

(b) any and all time lost by reason of discharge from inaccessible areas.

Tank tops, tunnels, brackets, bilges and all other similar provisions within the vessel's holds are to be properly protected by Owners, at their expense, against damage by grabs. Owners failing to provide such protection are to be responsible for all consequences arising therefrom. Any stevedore damage, or dispute arising therefrom to be settled directly between Owners and Stevedores and time used in repairing such damage shall not count as laytime.

This is a comprehensive clause and covers a number of important points, the first of which is the ability to discharge the cargo by grabs. It is clear (even though stated in line 2) that this charter party form is intended for single deck bulk carriers. Thus should the cargo be loaded in any spaces where a grab cannot be used, the additional time taken to discharge the cargo will not count as laytime. This would penalise any tweendecker in this trade.

The other point that arises is that grabs can do considerable damage to a ship's structure. The clause puts the responsibility on the owner for protecting vulnerable parts of the ship's holds. This is again likely to inhibit the use of a tweendecker in this trade. A bulk carrier would be designed with clear holds so it is unlikely that the sort of damage described in this clause could occur. With a tweendecker, however, the holds contain parts of the ship's structure that could easily be damaged. Finally in the event of grab damage being caused by stevedores, the owner must deal with the stevedores directly without involving the charterer or the receiver.

Another problem involves the stowage of the cargo and the trim of the ship. Again all the charter party forms have a clause that gives the master the responsibility for ensuring that the ship always has a safe trim for any voyages that are to be undertaken under this charter. This clause gives the master further authority, should it be needed, to require the stevedores to load the cargo under his direction in order that the ship will have a safe trim at all stages of the voyage.

Arbitration

All charter parties have an arbitration clause that covers how disputes are to be settled. The advantage of arbitration is that disputes can be settled more quickly and cheaply than by resort to the courts. There is an opportunity for the dispute to be heard by arbitrators with commercial experience rather than judges, and unless the dispute is to be heard in the USA the disputes can be settled in confidence. There are normally two main features of an "arbitration clause". First where it is to be held. This will decide under which country's law the arbitration is to be held. The main centres for arbitration are London and New York. The clause will also give a time limit after which all claims are barred and provide how arbitrators are to be appointed, typically one by each party and a third or umpire by the two chosen arbitrators. The owner should request that a clause is added requiring that the arbitrators are all shipping men, conversant with shipping matters, and, if the arbitration is to be held in London, are members of the LMAA (London Marine Arbitrators Association).

TIME CHARTERS

Time chartering forms the second method by which a ship owner can employ his ship. In this case by accepting the responsibility for finding the cargo and obtaining the fuel the charterer accepts more of the risk. The price for this additional security is that time chartering is not generally considered to be as profitable as voyage chartering. The time charter, however, will give a trader stability of costs against movements of freight rates during the period of the charter.

The clauses used in this discussion are taken from the New York Produce Exchange Form, which is a common time charter form in general use. It is considered to be strongly biased in favour of the charterer.

Many of the clauses in a charter party are common to both time and voyage charters. The main points of difference that are found in a time charter party are:

1. *Description of the ship.* This will be much more detailed than for a voyage charter, it is important that these are correct as they form the basis of contract and any claims for underperformance will be based on a comparison of actual performance against the details given in this description. The details required by the different charterers will vary depending on the trade envisaged but they are all likely to include the following:

   (a) *Numbers of holds* with their cubic capacity including details of the maximum weight tons (tonnes)/square ft (metre) that can be loaded on the tank tops. This must be taken into account if the charterer is contemplating the loading of high density cargoes such as iron ore or steel coils.

(b) A description of any *cargo handling equipment* (if any). This must be sufficient for the charterer to judge the suitability of the ship's equipment to load or discharge the type of cargo that they had in mind when chartering the ship.

(c) *Capacity of fuel oil tanks.* A large fuel capacity will enable the charterer to take advantage of cheap bunkers at different ports. The grade of bunkers for both main engine and generators will be agreed, and a statement may be included that the charterer will endeavour to supply fuel that meets the relevant British Standards (BSMA 100).

(d) *Speed and consumption.* Possible weather conditions (see later remarks under performance). Some charterers will also want to know the range of speeds and corresponding consumptions. This will enable the charterer to select the most profitable speed for a given voyage.

(e) *Dimensions length, breadth, draft, and air draft* (air draft gives the clearance for passing under bridges).

(f) *Certificates.* The owner will warrant that all the statutory certificates required for trading are, and will remain, in force for the duration of the charter.

(g) *United States Coast Guard certificates.* There are additional certificates regarding liability for oil pollution that must be carried on ships trading to the USA.

(h) *Date of last drydocking.* Any ship's speed depends on the state of the hull. After a period at sea it will be covered with marine growth (grass, barnacles, etc) which will reduce the speed.

(i) Whether the hull is painted with *SPP* (self polishing primer). These are special high quality paints that are polished by the action of the sea. This means that the speed does not drop off between dry dockings in the same way as conventional paints. This will give an indication as to whether the owners speed and consumption warranty is likely to be maintained over a period of time.

In addition the charterer may require hull plans, general arrangements plans, draft and displacement scales. These are required so that the charterer can assess the suitability of the ship for a proposed voyage.

2. *Period.* The period of the charter is stated usually 15 days more or less at charterers option. There will be a delivery date and a cancelling date with the same conditions as voyage charters. For trip charters the approximate duration of the voyage will be stated, for example about 45 days (see BIMCO report on length of voyages).

A clause goes on to state that any time used by the charterer after the expiry of the stated period will be paid for at the rate given in the charter party.

The charter party will state where delivery and redelivery is to take place:

(a) Dock or berth.

(b) Arrival at pilot station (APS).

(c) Dropping outward pilot (DOP).

(d) When passing a point of land (eg Gibraltar).

The charterer may agree to pay a *ballast* bonus if the ship has a ballast voyage to reach the delivery position or possibly a voyage to the next possible loading port. This will compensate the owner for the costs of the ballast voyage while protecting the interests of the charterer in the event of the ship not being ready to load by the cancelling date. Should the ship not reach the delivery port by the cancelling date, the charterer has the option to cancel the charter and is relieved of the obligation to pay a ballast bonus. Owners would want payment of the ballast bonus and bunkers on delivery with the first hire payment.

The division of expenses is more complicated in a time charter than with a voyage charter. It can be crudely stated that in a voyage charter the owner pays for everything except the cargo handling costs. With a time charter, a crude division would be that the owner pays everything connected with the running of the ship while the charterer pays all the expenses connected with the commercial operation of the vessel.

The NYPE form states:

1. The Owners shall provide and pay (when on hire is always added) for the insurance of the vessel and for all provisions, cabin, deck, engine-room and other necessary stores, including boiler water; shall pay for wages, consular shipping and discharging fees of the crew; shall maintain vessel's class and keep her in a thoroughly efficient state in hull, machinery and equipment for and during the service.

2. The Charterers, while the vessel is on hire, shall provide and pay for all the fuel except as otherwise agreed, port charges, pilotages, towages, agencies, commissions, consular charges (except those pertaining to individual crew members or flag of the vessel), and all other usual expenses except those stated in Clause 1, but when the vessel puts into a port for causes for which the vessel is responsible, then all such charges incurred shall be paid by the Owners. Fumigations ordered because of illness of the crew shall be for owner's account. Fumigations ordered because of cargoes carried or ports visited while vessel is employed under this Charter shall be for Charterer's account. All other fumigations shall be for Charterer's account after vessel has been on charter for a continuous period of six months or more.

Charterers shall provide necessary dunnage and shifting boards, also any extra fittings requisite for a special trade or an unusual cargo, but owners shall allow them to use any dunnage and shifting boards already aboard vessel.

Thus there is clearly a need to spell out all the additional costs that are likely to occur during the period of hire.

3. The following *surveys* are held before the ship goes on hire and on redelivery. The purpose of the surveys is to ascertain:

(a) Bunkers on board. The charterer pays for the bunkers on board the ship, and possibly boiler water on delivery. The owner pays for the bunkers remaining on redelivery. The charterer should provide that the vessel has sufficient bunkers for her to reach the next main bunkering port.

(b) Condition of the ship. When going on hire the surveyor will establish that the description of the ship is accurate. He may want to see that the equipment actually works. The charterer is responsible for certain damage to the ship during the period of the charter; for example damage caused by cargo handling or due to unsafe ports. The surveyor will establish the extent of any unrepaired damage.

(c) Condition of holds. The surveyor will check that the holds are fit for the cargo before going on hire. Note that hold cleaning after the first and subsequent cargoes is at the charterer's expense. When the ship goes off hire the charterer is liable to clean the holds, frequently the charter party will state "shovel clean", ie the holds are cleared with shovels and no further cleaning will be carried out. This would represent an additional cleaning expense for the shipowner if he wishes to charter the ship for a cargo that requires clean holds such as grain.

4. *Cargo Exclusions*. This clause will state whether any cargoes are to be excluded, possibly sulphur, steel coils or any other cargo likely to damage the ship. Common wording is "any cargo harmful, injurious or damaging".

5. *Trading limits*. In most cases these will be drawn to cover most of the world. However the owner will exclude trading outside the Institute warranty limits, this is the area in which the normal hull insurance applies. Should the charterer want the ship to trade outside these limits, and the owner agrees, the charterer would have to pay for any extra insurance premium and other costs involved. In the case of small ships there may be a need to restrict trading to certain areas because the rules of the flag state regarding crewing will not allow the ship to trade outside these limits.

The owner may wish to exclude certain countries in which he does not want his ship to trade—for example, at the present time, Iraq and Iran; or, because of the political situation, Sweden and Australia because the ITF is active in these countries and if the ship has a "flag of convenience" she might be blocked by union action and the owner forced to pay large sums of money to the crew in order to get the ship released. The owner may also want to exclude Israel and South Africa because of possible embargoes imposed by various Arab countries and Nigeria, although in the case of South Africa, it is unlikely that a charterer looking for a ship to carry coal would accept such a restriction.

7. *Payment of Hire*: Payment of hire shall be made so as to be received by Owners or their designated payee in New York, ie   in United States Currency, in funds available to the Owners on the due date, semi monthly in advance, and for the last half month or part of the same the approximate amount of hire, and should the same not cover the actual time, hire shall be paid in for the balance day by day as it becomes due, if so required by owners. Failing the punctual and regular payment of the hire, or on any breach of this Charter, the Owners shall be at liberty to withdraw the vessel from the service of the Charterers without prejudice to any claims they (the Owners) may otherwise have on the Charterers.

Time shall count from 7 am on the working day following that on which written notice of readiness has been given to the Charterers or their agents before 4 pm but if required by the charterers, they shall have the privilege of using the vessel at once, in which case the vessel will be on hire from the commencement of work.

This clause allows the owner to withdraw the ship after giving notice in the event of hire not being paid. In practice, the withdrawal can be more complicated than this clause leads one to believe. There is a procedure to be followed through and BIMCO strongly advise an owner to consult their P & I Club before contemplating such a move. There is always the additional problem of ownership of any cargo belonging to a third party, which may be on board at the time when the owner wants to withdraw the ship.

A shift in freight rates on the voyage market can lead to one or the other party looking for a pretext to end a time charter. Should rates on the voyage market fall after the charter is fixed, the charterer may well be looking for a pretext to rid himself of what has become an unprofitable contract, while it is obviously in the owner's interests to ensure that no pretext is given for the charterer to escape from the contract, otherwise he is going to find his ship on the spot market with diminished prospects of profitable employment. Should the converse apply and freight rates rise substantially, the owner will be looking for a pretext to return to what is now a more profitable spot market and it becomes important that the charterer ensures that he does not provide an opening. This can lead to a cat and mouse game between the owner and charterer with one or other of the parties looking for a pretext to get rid of what has become an unprofitable contract. Thus, when looked at from this point of view, what might, under normal circumstances, be considered to be rather trivial disputes, become exaggerated.

Lien: The owners shall have a lien on all cargoes and all sub freights for any amounts due under this Charter including general average contributions, and Charterers shall have a lien on the vessel for all moneys paid in advance and any overpaid hire or excess deposit to be paid at once. Charterers will not suffer, nor permit to be continued, any lien or encumbrance incurred by them or their agents, which might have priority over the title and interest of the owners in the vessel.

The purpose of this clause is to provide both the owners and charterers security in a situation where the other party defaults in some way. Despite the ability to withdraw the ship in the event of late payment of hire under a time charter, there is always the possibility that the owner might have to complete the voyage in order to deliver a cargo before being able to withdraw from the time charter. This clause ensures that in such an event, the owner can claim any monies due from the cargo owner or from the freight. It is also important that in the event of a charterer being in financial difficulty he is prevented from using the ship as a security for money advanced, possibly to pay port costs bunkers or some other cost which could place an owner in jeopardy.

The lien in freight is easier to exercise than a lien on the cargo, which is frequently the only security available to an owner under a voyage charter.

Performance is an area of time chartering which probably is the cause of most disputes. The clause in the NYPE charter party reads:

In the event of the loss of time from deficiency and/or default of officers or crew or deficiency of stores, fire, breakdown of or damages to, hull, machinery or equipment, grounding, detention by average accidents to ship or cargo unless resulting from inherent vice, quality or defect of the cargo, drydocking for the purpose of examination or painting bottom, or by any other similar cause preventing the full working of the vessel, the payment of hire and overtime, if any shall cease for the time thereby lost. Should the vessel deviate or put back during a voyage, contrary to the orders or directions of the Charterers for any reason other than by accident to the cargo, the hire is to be suspended from the time of her deviating or putting back until she is in the same or equidistant position from the destination and the voyage resumed therefrom. All fuel used by the vessel when off hire shall be for Owner's account. In the event of the vessel being driven into port or to anchorage through stress of weather, trading to shallow harbours or to rivers or ports with bars, any detention of the vessel and/or expenses resulting from such detention shall be for the Charterer's account. If upon the voyage the speed be reduced by defect in, or breakdown of, her hull machinery or equipment, the time so lost, and the cost of any extra fuel consumed in consequence therefore, and all extra expenses shall be deducted from the hire.

This off hire clause is one of the crucial clauses in any time charter because it defines the period when the charterer is not liable to pay hire. The clause is comprehensive, covering any situation where the vessel is not meeting the performance requirements laid down in the charter party document. Whenever the charterer considers that the ship is not performing as per the charter party, the charterer is free to deduct hire in respect of the time that he considers that he has lost. This can be costly for a shipowner, in order to challenge such a deduction it may be necessary to resort to arbitration.

This clause must be read in conjunction with the description of the vessel on the first page of the charter party. It is this description that is the basis of the contract. The description will include the speed and consumption and should the ship not maintain this speed, then according to this clause the charterer is free to deduct the hire for any extra time taken on passages resulting from the reduced speed. Charterers frequently employ companies such as Ocean Routes to monitor the ship's performance and substantiate any claims for underperformance. The charter party will state whether performance is to be monitored by Ocean Routes or from the ship's logbooks. There is no mention of a bonus for the owner should the vessel exceed the time charter requirements, so there is the temptation for the owner to exaggerate the vessel's speed in order to gain a slightly higher rate. In practice the vessel might find it difficult to maintain such a speed on a day-in day-out basis.

It should be noted that any ship's speed depends on the state of the hull and the weather, which is why the date of the last drydocking is frequently stated in time charterers and there will be reference to the ship coming off hire for drydocking and hull cleaning. Many ships only drydock about once every four years. This is made possible because the classification societies will allow the various inspections to be carried out while the ship is afloat. It is also possible to clean the hull without drydocking by using divers equipped with special scrubbing equipment. Thus it is possible for a ship to maintain her charter speed without the expensive off hire time needed for drydocking.

The clause does not make any mention about weather. When the weather is bad it would not be reasonable to expect a ship to maintain her full speed. Even if the ship had sufficient power to maintain full speed in all weathers she would suffer considerable damage if forced to proceed too fast in stormy conditions. It is frequently accepted in time charters that during periods when the wind speed exceeds Beaufort force 5, the time should not count when monitoring the ship's performance. Another factor that should be considered is the need to proceed at a "moderate" speed when in fog, the "rules for prevention of collisions at sea" are clear on this point, hence the need to include under speed the words *"weather and safe navigation permitting"* when describing a ship's speed in the charter party.

Difficulties can occur when the wrong grade of fuel is supplied. There is a British Standard for fuel oil and the owner should insert a requirement that only bunkers meeting this specification are supplied—in addition to the ship not being able to meet the speed and consumption requirements, there is a possibility of damage to the machinery. Hence the reference to BSMA 100 specification for fuel oil in the ship's description.

The charterer is also allowed to ask the owner to replace the captain and officers should he consider that they are not adequately representing his interests.

Thus although the time charter would appear to provide an advantage to the owner in providing the security of long-term employment, in practice the owner has a number of obligations that he must meet if he is to avoid expensive claims which will turn what appeared to be a profitable contract into a financial disaster.

Chartering is a complicated business. In the short space available it has only been possible to give a brief outline of some of the pitfalls involved when a ship is chartered for a coal cargo. Should readers require further information they are advised to consult some of the many specialised books written on the subject.

REFERENCES FOR CHAPTER 8

For further information on chartering, the reader should consult:

Lars Gorton, Rolf Ihre and Arne Sandevarn *Shipbroking and Chartering Practice* (3rd ed., 1990, Lloyd's of London Press).
BIMCO *Check Before Fixing.*
Gram on Chartering Documents (2nd ed. 1988, Lloyd's of London Press).
William Packard, *Sea-Trading* Vol 3 "Trading".

PART III

Origins, properties, utilisation and evaluation of coal

This part of the book is split into two sections. The first section starts with understanding of the role of coal in the present economic spectrum, the origins of coals and their various characteristics and a look at principal utilisations of coal.

In the course of this study it will be clear that detailed technical procedures are required in order to ensure that a coal is suitable for a given industrial process.

The second section looks at these technical procedures from the standpoint of a professional cargo inspector and analyst, highlighting some of the operational problems that can arise in the course of evaluation.

If this particular approach enables the reader to appreciate that the evaluation of a material as seemingly innocuous as coal is a many faceted venture, with many pitfalls and difficulties, and is not something to be taken for granted, then the objective of the writer will have been achieved.

I would like to express my great appreciation to my good friend and associate James Docherty, General Manager, Minerals and Water of Cargo Superintendents Co (A/SIA) Pty Limited, Australia who was kind enough to read the proofs of this contribution, as well as add some invaluable and in-depth examples and opinions which serve to add authenticity and topicality to the text. He also donated the photographs of laboratory equipment. His kindness and generosity are much appreciated.

SECTION 1

Origins, properties and utilisation of coal

The logical starting point of such a review is to examine the nature and origin of coal itself. This builds up a picture of the nature of the material to be evaluated, and the nature of the tests involved obviously relates to the composition of the material in question. That composition naturally relates to the source and origin of the coal.

The only point of performing elaborate tests on any raw material is to ascertain whether it is suitable for its end-use. Consequently, any tests on a coal will relate to the particular use to which that coal is to be put.

This review will try to present a complete picture of coal as a raw material tracing its handling from the moment it is dug out of the ground to the time when it is subject to industrial use.

CHAPTER 9

Origins

FORMATION

Coal is an impure form of carbon, occurring naturally as a stratified sedimentary rock. It is usually black in colour, although it can occur as a dark brown mineral. The process of the formation of coal began several hundreds of millions of years ago in what geologists term the carboniferous period.

The actual process of forming coal probably took about 60 million years. It is, in effect, a compact stratified mass of decayed plants, which have been converted chemically and physically into a mineral deposit.

The original process of coal formation took place in areas in which there had been luxurious vegetation and especially a preponderance of trees. The decaying process was catalysed by bacteria and the overall conversion from organic plants was also effected by the heat and pressure from overlying rock formations. The chain of conversion begins with the conversion of plant debris into peat bogs and the process from then on passes from lignite to brown coal to sub-bituminous, bituminous and finally to anthracite.

The conversion process from the decay of organic plants through the stages of peat formation to anthracite, is known as "coalification"; a clumsy word but nevertheless descriptive. The degree of coalification is described by the concept of "rank".

In essence, rank indicates the extent of alteration reached by the coalification process. Thus, lignite and sub-bituminous coals would be considered low-rank whereas anthracite is high-rank. This progressive series reflects the increase in carbon content, decrease in volatiles and to some extent increase in calorific value. The range of coals from bituminous through to anthracite is termed humic coals.

Some coal fields can have as many as 100 seams of coal, and this illustrates that the process of growth, decay, submergence and silting must have occurred many times. Some seams can be as little as 5 cm in depth while on the contrary others can be up to 100 m. The average thickness is in the order of 2–3 m.

As the heat and pressure referred to above squeezed water and gases out of the mineral body, the peat beds first became converted to lignite, which is a very soft, really half-formed coal usually brown in colour. The next stage was the conversion into bituminous coal, this being the most commonly known form to the man in the street. The final stage was the conversion into anthracite, the hardest form of coal, which is very nearly pure carbon.

MINING

Obviously the method used to mine coal depends principally on the depth of the strata of coal beneath the surface.

In many of the largest producing countries such as Australia, South Africa, USA and USSR, the main deposits are relatively close to the surface and once the overburden (soil and light rock) has been removed, the coal can then be mined in an open-cast method, which effectively involves removing the coal by mechanical diggers. In many of the older coal fields in Europe, such as the UK, Poland and Germany, the seams are often very deep (sometimes up to 2000 m below ground level) and therefore the coal has to be mined by underground methods involving the sinking of many deep shafts.

COMPOSITION

The composition of coal depends on the location of the deposit and the origins of the organic matter that have led to its formation. Also, it depends upon the conditions of temperature and pressure which existed during transformation, as well as the depth of the coal layers.

The coal extracted from the coal seam or seams is usually referred to as "run of mine" (ROM) coal and is a mixture of coal and "dirt bands" associated with the mining conditions from which it is extracted. ROM coal is normally of low quality and in many cases needs some improvements before it can be utilised. In many coal-fired power stations, ROM coals are used directly without improvement. The composition of this ROM coal determines what types of coal preparation processes will be required to improve the product coal from the ROM stage. Many processes are used including jig washers, dense media cyclones, froth flotation etc. For instance, Australia—the largest exporter of coal in the world—exports predominantly washed coal but mainly uses ROM coal for power generation internally.

Coal composition is very important in determining its utilisation and potential for export. Many coals lend themselves to be easily cleaned and have certain characteristics that are sought after in the international coal trade. However, other coals are the opposite, difficult to clean, awkward in utilisation and difficult to sell internationally.

GENERAL CLASSIFICATION

In Europe probably the best known classification based solely on analytical data is the Seyler Classification. The analysis data are obtained on a dry, ash-free basis with the main parameters of carbon and hydrogen. It can be seen that most coals classified in such a way occupy positions within a clearly defined curved band. Also, volatile matter and calorific value results can be used in coal classification. The Seyler system works well for many coals, outlining their coking coal potential. However, many low vitrinite percentage coals, whilst having a good coking potential, do not indicate this, using the Seyler Classification technique. Commercially Seyler is little used by customers these days as many coal-producing countries have developed systems for classifying coals on a national basis. These can be readily found in published coal standards.

CHAPTER 10

The petrographic classification of coal

Petrography is the microscopic examination, description and analysis of rocks and minerals. Petrology uses the petrographic results plus any other relevant analyses to interpret factors such as origin, method of formation, possible uses and specific applications of the material being studied.

Most rock and mineral petrography requires thin section, transmitted light microscopy. While this technique is appropriate to some coal studies, the effort and cost of suitable coal sections is prohibitive for routine industrial application.

Most coal petrography and all commercial/industrial coal petrography uses incident/reflected light microscopy similar to metallurgical microscopy.

Coal specimens or samples may be examined as either polished lumps or polished grainmounts. Polished lumps are suitable for some geological studies. However, most coal samples are prepared by mixing a crushed sample with a binder resin, allowing the mix to set in a suitable mould, then cutting and polishing the block.

A petrographic analysis is made by a point count of the polished surface.

The microscope requirements and procedures for sample preparation and coal petrographic analysis are described in various national standards, most of which are based on ISO7404.

In hand specimen coal is made up of alternating dull and bright bands and lenses of varying thickness. The dull bands may be termed durain, the brighter bands vitrain and the intimately mixed/thinly layered bands clarain. Shale bands may also be present.

Microscopic examination shows the coal bands to be made up of many different structured and unstructured entities. (In younger coals, some structural features can be identified as related to presently existing plants; older coals are related to and formed from extinct plant species.) These microscopic entities are termed *maceral*; analogous to the minerals forming rocks.

While the macerals may be regarded as the building blocks of coal, they are unlike inorganic minerals, which generally have crystalline structures and a fixed or narrow range of chemical composition. The macerals are organic with no clearly defined crystal structure and a composition (carbon, hydrogen, nitrogen and oxygen) that varies with the coal rank.

Related macerals belong to one of the three maceral groups—*vitrinite*, *liptinite* (or *exinite*) and *inertinite*.

For bituminous coals, the main *vitrinite group* macerals are *telocollinite*, the major component in bright coal or vitrain, and *desmocollinite*, which forms much of the ground mass in dull coal.

The *inertinite group* includes structural macerals fusinite and semi-fusinite, unstructured lenses termed macrinite, fine grained micrinite, and fragments of other inertinite macerals termed inertodetrinite.

The *liptinite group*, which rarely exceeds 5% of a coal, includes the macerals resinite, sporininite and cutinite plus other less common macerals.

The same maceral groups are recognised in lower rank (brown coal) coals and anthracites but individual macerals exhibit different appearance.

Coal rank is the position of a coal in the series brown coal—bituminous coal—anthracite. It can be readily determined petrographically by measuring the reflectances of the vitrinite group macerals.

Traditional methods—carbon content, volatile matter—have the disadvantage in that, while they vary with coal rank, they are also influenced by variations in maceral composition. An inertinite rich coal will have higher carbon and lower volatile matter than a vitrinite rich, bright coal of the same rank.

Procedures for vitrinite reflectance measurement are covered in various natural standards, generally based on ISO7604.

Table 81 Vitrinite reflectance: (% measured in oil, light wavelength 546 nm)

Lignite	<0.2
Sub-Bituminous	0.2–0.4
High Volatile Bituminous	0.4–0.8
Medium Volatile Bituminous	0.8–1.3
Low Volatile Bituminous	1.3–2.2
Anthracite	>2.2

Coal petrography is of value in all areas of coal assessment and quality monitoring.

For an exploration geologist, petrography of weathered outcrops and chip samples from non-cored boreholes identifies the coal rank, relative proportions of the macerals present, as well as the nature and distribution of minerals in the coal. Petrographic analyses help to correlate and evaluate the detailed laboratory chemical and carbonisation tests undertaken on borehole core samples.

Similarly for operating mines, petrographic analysis of routine samples from development headings helps to identify changes in coal seam character, either variations in rank, maceral composition or mineral content, that affect coal quality form the mine.

Petrography is useful in coal preparation plants both as an aid in trouble shooting and solving quality variation problems and in occasionally monitoring the performance of the various washery units.

For many coals, the coal preparation processes of crushing and screening segregate coal fractions with significantly different maceral compositions. How these fractions are treated and combined determines the quality of the final products.

Coal buyers can benefit from the use of petrography to confirm that the coal contracted is actually supplied—it is not difficult for a petrographer to identify two or more component coals in a sample, but it may not be readily obvious from simple proximate analysis.

The above classifications are shown schematically in Figure 10.

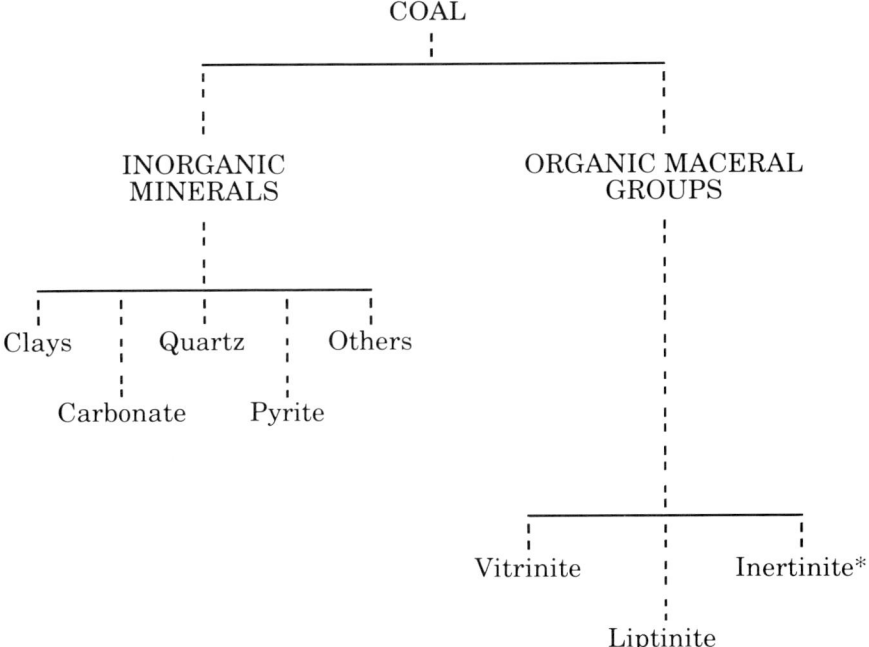

Figure 10 Coal Petrography.

APPLICATION TO COKING COAL

For a coal to be suitable as a major component in a coking blend, it must have certain caking properties when heated. It must melt, develop a minimum fluidity and resolidify to a porous mass with a minimum compressing strength and resistance to abrasion.

Only coals of bituminous rank are suitable and their relative "coking behaviour" varies significantly with changes in rank and maceral composition.

It should be noted that actual coking conditions, size of grind, heating rate, bulk density, oil addition, soaking time etc, also ash content of coal, weathered coal, etc, have an equally important effect on the actual coke formed.

However, in general terms, it is the vitrinite and liptinite macerals that play the dominant role in the coking process. Both become fluid on heating and resolidify to bind the coke mass. These macerals are no longer recognisable in the coke.

The inertinite macerals in general do not melt or become fluid on heating and tend to form the "aggregate" in the coke mass. Much of the inertinite is still recognisable after coking. The finer inertodetrinite is generally well incorporated into the coke cell walls, tending to thicken and strengthen them. Larger fragments of semifusinite or fusinite may not be well bonded into the coke, particularly if the original vitrinite content was low.

Various empirical formulae have been developed relating petrographic analyses to coke strength. Most are based on American publications of the early 1960s.

The terms "composition balance index" and "strength index" were introduced by Shapiro-Gray and Eusner in 1961. The composition balance index (CBI) is the amount of "coking reactive" of a particular rank coal required by the "coking inert" component to make the optimum strength coke for that particular coal rank. The strength index (SI) is an empirical measure of the relative strength of coke from coal of a particular rank, taking into account the amount of "coking inert" present in that coal.

The "coking reactive" components generally considered as vitrinite + liptinite + part of the inertinite (varies from 33–45% depending on the coal).

The "coking inert" component is 100%—reactive component.

Published tables are used to calculate CBI + SI for a coal or coal blend using total reactives and inerts calculated from the petrographic maceral composition and vitrinite reflectance distribution. The calculated CBI and SI are then plotted on printed graphs to give the predicted coke strength, either ASTM stability index, or other coke indices, for example micum, JIS, etc.

Coke strength prediction tends to work best for coals that are known or used by the coal technologist making the predictions. Unknown coals should be tested in pilot coke ovens to check the correlation factors used.

There is considerable on-going research into coke petrography that should help improve predictions of coke strength and coke reactivity from coal petrographic data. There is increasing interest in coal petrography as an aid to assessing fuel coals and their behaviour in pulverised coal-fired boilers.

Combustion rates and char characteristics vary with different maceral composition and coal rank.

Coal liquefaction and gasification are other areas in which petrography can play a part in investigation and research.

In summary, coal petrography has an important place in routine programmes to monitor coal quality during production, treatment and end-use. It is also invaluable in any coal exploration or research project.

THE PRODUCTION OF COKE

It is clear from the foregoing that having petrographic data on given coals enables judgements to be made for blending in order to produce coke of the required characteristics for a given industrial use.

Coke is the product of heating coal in a so-called coke oven in the absence of air. The process is termed carbonisation. Usually it is a blend of coking coals that are carbonised and the component coals are selected in order to produce the properties required to make a coke suitable for the industrial process in question. The Japanese, for example, have been known to blend up to 20 different coals in order to produce a range of highly specific coke products. Coke ovens are usually arranged in a battery form (ie side by side), and fed by crushed blended coal usually 80 % of which is finer than 3 mm top-size.

The coking coal blend is fed into the coke ovens and heated to a temperature of 1,100°C over a period of 12—24 hours.

When carbonisation is complete, the still-incandescent coke is ejected from the oven and quenched usually with water or inert gas. The latter process is now becoming more common because of the savings in energy due to heat conservation. The use of inert gas is called dry quenching.

The products of carbonisation are coke (60–70%); coke breeze (5—10%); together with a mixture of coke oven gas, tar, ammoniacal liquor and other ammonium products, and light oil. The various qualities of the coking coal blend that have to be met in order to provide coke of suitable quality for use in a blast furnace are as follows:

Ash: normally 8.5% maximum—if ash content too high, it will reduce the efficiency of the blast furnace.

Volatiles: 20–35%—if volatile matter too high, the coke yield from carbonisation is reduced.

Total Moisture: 10%—if moisture too high, the amount of carbon available is reduced and there are also difficulties in handling.

Elements: such as sulphur, phosphorus, sodium and potassium are of critical importance due to their effect on either the steel product or coke. For example, if there are excessive amounts (say over 3 % in the ash) of sodium or potassium oxides, then the reactivity of the coke will be too high. There are also strict controls on the phosphorus levels in both iron ore and coke due to the effect of making steel too brittle. Sulphur, likewise has a deleterious effect on steel.

In addition there are a number of tests specifically designed for coking coal. Earlier, the caking properties were highlighted and the most simple, unsophisticated test involves the estimation of the "crucible swelling number" (CSN)—also known in the USA as the "free swelling index". Quite simply, crushed coal is heated rapidly in a crucible and the resultant coke button obtained on cooling, is compared with a standard series ranging from 0 (no caking) through to 9 with superior qualities graded 9 +. Such a simplistic test is vulnerable to factors such as size and moisture content of the sample as well as oxidation characteristics.

An alternative to the CSN is the Roga Index. Here, the crushed coal sample is mixed with anthracite and heated to produce the coke button. This is then submitted to a drum test, a procedure for establishing the strength properties of coke. A specially designed drum is rotated and the coke inside is lifted and then dropped to the bottom of the drum during rotation. Drum indices are assigned from formulae involving the size analysis of the resultant material. The Roga Index is based specifically on the − 1mm coke, also being calculated from a formula.

The most commonly used caking test in Europe uses the Audibert-Arnu dilatometer. This measures the expanding and contracting characteristics by compressing finely crushed coal into a pencil-shaped sample and measuring the maximum expansion (or dilation) and maximum contraction during a carefully controlled, slow heating process. The temperature at which the coal initially softens (and so contracts) is recorded and then, as it becomes plastic in nature, the maximum contraction and dilation are recorded and expressed as percentages of the original pencil length. Predictably, this test is very sensitive to the oxidation characteristics of the coal. The other dilatometer test—the Ruhr—differs mainly in the degree of compaction of the coal pencil sample. As a result, Ruhr results show a lower contraction and higher dilation than those from Audibert-Arnu tests.

In Japan and the USA, the "Gieseler plastometer" test is widely used to measure fluidity characteristics of coking coal. This test records the plastic range of fine coal whilst it is heated slowly and the fluidity throughout that range is measured. The final result reported is the maximum fluidity. This test is also extremely sensitive to oxidation.

CHAPTER 11

The industrial utilisation of coal

The three principal industries that utilise coal are:

1. Steel.
2. Power.
3. Cement.

In describing the essential processes of these industries it will become clear which properties of coal and coal products are of key importance. These properties can, in turn, be related to the critical components that each industry requires to be evaluated.

STEEL INDUSTRY

The main raw material used in the steel industry is iron ore. This is generally available in three principal forms: fines, lump and pellets. Lump ore and pellets can be charged directly to a blast furnace, whereas fines are mainly used as sinter feed. The resultant sinter, in turn, provides additional furnace feed.

The blast furnace is the classic production process for the reduction of iron ores. The furnace is charged through the top with weighed amounts of coke, limestone, sinter and iron ore. The actual thermodynamic and chemical kinetic processes involved are very complex. One of the most essential aspects of the process is the control of the material balance of the feed. Blast furnaces have to operate continuously, so just as the feed must be continuously balanced, so must the removal of the products, both gaseous (oxygen, nitrogen, carbon monoxide and carbon dioxide) and liquid (molten pig iron and slags).

The basic feed for a blast furnace comprises iron ore, coke of suitable composition for blast furnace use, scrap and a charged flux. The materials are charged into the top of the furnace and molten iron and molten slag are tapped off from the bottom of the furnace. Hot air is blasted into the lower part of the furnace and gas with low calorific value recovered from the top.

The iron ore is reduced by carbon monoxide at an ambient furnace temperature of approximately 1,650°C. Additional fuel in the form of fuel oil or tar or pulverised coal can be injected into the furnace together with the blast air in order to maintain a stable temperature. If pulverised coal is injected into the blast furnace instead of coke, the coking coal can be replaced by steam coal. The usual ratios of blast furnace feed are:

ORE : COKE : SCRAP : FLUX
100 : 33 :   7   :   1

(These are orders of magnitude and not a precise composition)

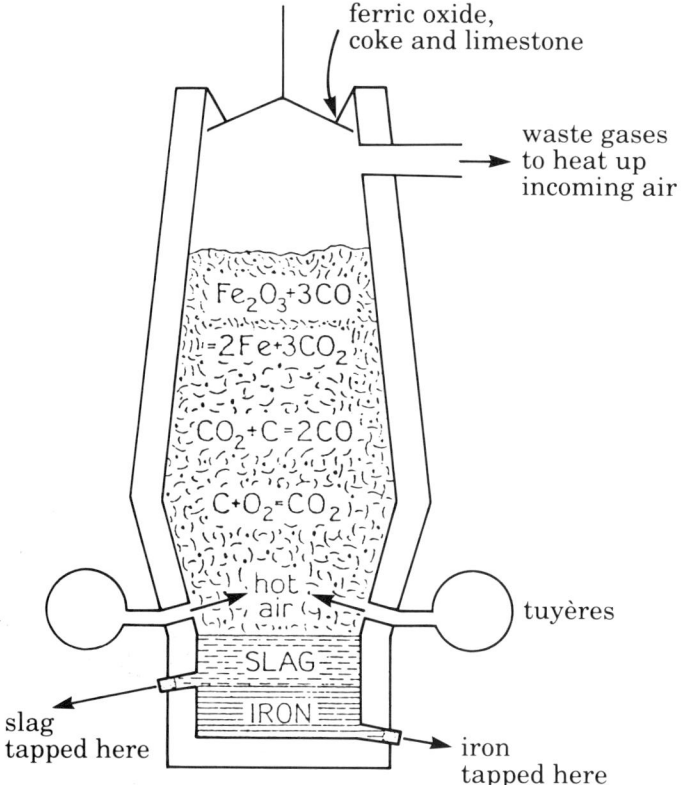

ferric oxide,
coke and limestone

waste gases
to heat up
incoming air

$Fe_2O_3 + 3CO$

$= 2Fe + 3CO_2$

$CO_2 + C = 2CO$

$C + O_2 = CO_2$

hot
air

tuyères

SLAG

IRON

slag
tapped here

iron
tapped here

Figure 11 Blast-furnace.

It is essential to control the impurities that can enter the iron. For example, carburisation of iron occurs at temperatures in excess of 1,000°C with the formation of the carbide, Fe_3C. Carbon will reduce silica at temperatures in excess of 1,500°C so that ferrosilicon enters the molten iron in an amount proportional to temperature and the basicity of the slag.

Phosphorus enters the iron as Fe_3P after high temperature reduction of phosphates by which it is introduced. As this makes iron brittle its presence must be strictly controlled.

Sulphur enters by way of the coke and initially the sulphide FeS is formed. However, desulphurisation occurs by way of slag formation:

$$FeS + CaO + C \longrightarrow CaS + Fe + CO$$
$$\qquad \text{coke} \qquad\qquad\quad \text{slag}$$

The slag is principally calcium silicate ($CaSiO_3$) and this seems to act as a solvent to remove materials such as CaS from the system.

As well as the materials balance mentioned above, there is also a heat balance between input and output. One of the essential input features is the heat of fuel combustion used, making quality control, discussed earlier, so important.

POWER GENERATION

The basic principle of coal-fired power stations is to use coal to produce steam at high temperature and pressure, which will drive turbines in order to generate electricity. This is not a highly efficient process and only approximately 30% of the available energy in the coal is directly converted into electricity. The remainder of the energy is generally wasted either into the atmosphere from the cooling towers or water-cooling systems. Consequently, a 600 megawatt power station requires over 1,800 megawatts of thermal power to generate the electricity.

The coal used in power stations is usually pulverised so that, depending upon the rank of the coal, between 60 and 90% of the coal used will be of minus 75 microns (200 mesh) in size.

Coal this fine does not require a supporting grate and it can be used in very large boilers. In effect, the powdered coal is suspended as a cloud of fine particles in the combustion air. Therefore, the amount of time required for conversion into heat is much shorter than in ordinary stokers and more heat is released per unit volume.

In view of the handling problems associated with moving large amounts of very fine materials, it is clear that all power stations must have their own pulverisation plants in order to produce feed of the required size. Impact mills pulverise the feed coal, which on acquiring the requisite mesh size, is fed pneumatically to burners in the walls of furnaces. The pulverising mill is flushed with air which effectively drives the coal through the mill to the burners. This air is termed primary air which constitutes 20–25% of the total combustion air. The primary air is usually at a temperature in the range of 250–350°C, so that effectively the coal is dried whilst it is being pulverised and when it leaves the milling area, the coal is at an approximate temperature of 85°C.

The so-called secondary air is injected around each burner in order to enhance the combustion close to the tip of the burner.

The function of the burners is to provide a means of introducing the pulverised coal and the combustion air into the furnace in such a way as to produce a stable flame-front some distance from the burners. This would effect the efficient and speedy mixing of air and coal as well as directing incoming gases, thus utilising the full extent of the furnace chamber. The object of establishing the flame-front at some distance from the burner is to ensure that the flame does not impinge on the furnace wall.

As mentioned above, the alternative to pulverised fuel burners is called a stoker. This is used for feeding coal into a furnace and it is distributed over a grate. It is then subjected to combustion in a stream of air. Basically a stoker consists of a feed and a grate. They are categorised according to the manner in which the coal is fed onto the grate:

1. Under-feed stokers;
2. Over-feed stokers;
3. Spreader stokers.

Both pulverised fuel combustion and stoker processes have certain prerequisites for the coal used as feed. Detailed specifications are beyond the scope of this book, but it is of interest to look at some of the factors associated with the individual coal characteristics:

Ash content: Coal in itself does not contain ash. Ash is the material left after combustion. The higher the ash the less carbon and volatile are available to produce heat. Generally the higher the ash level the lower the calorific value of the coal becomes.

Total Moisture (TM): Total Moisture is a major factor in determining the heat value of the coal. Most calorific value results are expressed on an "as received basis" so that, in essence, the higher the Total Moisture of the coal, the lower the calorific value will be. Also, if the coal is too dry, dust can become a serious environmental problem. Many modern specifications allow for around 7% total moisture on export shipments as being an acceptable level. Climatic and coal washing conditions cause this figure to fluctuate markedly.

Hardgrove grindability index (HGI): This is a measure of the actual hardness of the coal and gives an indication of how much wear may be associated with handling this coal. If the HGI is below a level of 45 it may require more energy to crush it compared to coal at a level of 55. Therefore, HGI is an important factor when purchasing coal for electricity production.

Sulphur, nitrogen, etc: For obvious reasons there have to be maximum values set upon these elements due to pollution factors. Normally there will be local emission regulations to control the highest acceptable value of pollution.

Crucible swelling number: Usually a maximum of three is acceptable as high swelling coals tend to produce uneven combustion.

CEMENT PRODUCTION

The production of cement is highly energy intensive requiring heat for clinker production and electricity for grinding the feed and clinker.

There are two basic processes: wet and dry. As the names suggest, they relate to the condition in which the component-mix is ground and introduced into the kiln. In the wet process, the mix is introduced as a slurry with moisture in the range of 20–40%. The reduction of the moisture content improves the fuel efficiency of the plant.

For the dry process, the ground mix is introduced dry into the kiln. Dry process kilns are generally shorter as there is obviously no need to remove the slurry moisture by evaporation.

The basic cementation process is:

1. Basic raw materials (calcium carbonate, alumina, iron oxides, silica and other required components) are ground and blended together to form a uniform mix.

2. The blended mix is introduced into a kiln in order to remove all moisture.

3. The mix is calcined at 800°C and during this process carbon dioxide is evolved.

4. The temperature is increased to 1,400°C at which temperature final clinkering occurs.

5. The clinker is cooled, ground and finally mixed with approximately 5% gypsum to give the composition of the final mix the added advantage that the ash can be used as part of the product.

In view of the simple requirements for such a process, there are very few restrictions or specifications on coal that can be used for cement production.

The evaluation of coals

CHAPTER 12

Theoretical and practical considerations

In the course of this review, efforts will be made to evaluate the various options that exist when it comes to testing the material, as well as pointing out some of the hazards that have to be overcome on the way.

Space does not permit this to be a highly technical and detailed treatise; the intention is to introduce sufficient detail to enable the reader to appreciate the extent of the issue under discussion.

The world is currently preoccupied with the desirable balance that should exist between energy consumption and resources, as well as the environmental consequences of using fossil fuels. Whilst one cannot and should not minimise the current environmental concern, it is a fact of life that the modern world is "energy-hungry" and the influence of oil supply factors on the Gulf War underlines the political importance of sources of energy.

Coal remains not only the largest current source of fossil fuel but also the largest long-term reserve. Commercially, coal has traditionally been sold principally on an FOB (free on board) basis. This means that quality and quantity evaluations are performed at load ports. When coal was priced at under US$15 per tonne and the average size of cargoes was less than 10,000 tonnes, it did not make commercial sense to employ detailed and thorough weighing and sampling operations as the cost of such activities would have consumed a large part of the operating profits of those parties involved. However, with coal currently in mid-1991 at US$45 per tonne and shipments generally in excess of 60,000 tonnes, the value of cargoes is probably on average fifty times higher than at the end of the 1950s. The result is that any bias in the sampling process now costs someone money; in extreme cases, a lot of money. Therefore, anyone buying a large cargo of coal will usually insist on a thorough independent appraisal of weighing and sampling operations as well as the resultant chemical analysis in order to minimise the risk of severe financial losses.

All this means that over the last 15 years, suppliers and producers, independent inspection companies, consultants, mathematicians and other interested parties have been establishing sampling methods that can hopefully be applied universally so that a "consensus evaluation" within acceptable tolerances can be obtained. Obviously, contractually it is still the load port data that govern; but the buyer will often have checks made on arrival at the plant in order to confirm that the cargo conforms to the quality criteria required for the process in question. Whether it be steam coal for a power station or coking coal for a steel works, it is this confidence factor that is so vital to the establishment of long-term business relationships.

The creation of reliable sampling procedures cannot be overstated. However elaborate and sophisticated the procedures and equipment for sample preparation and analysis, if the original sample is unreliable, unrepresentative or biased, the damage is irreparable. Later on, statistical aspects will be considered, but suffice to say that in terms of the component parts of the overall precision of coal evaluation, sampling accounts for approximately 80%, sample preparation 10% and analysis 10% of the total.

Quite clearly, commercial contracts should, where feasible, stipulate the sampling procedures to be adopted. In turn, these should not be highly theoretical or too idealised, but represent operations that can be carried out with the equipment and facilities available. This is a point sometimes overlooked and resulting in either hurried improvisations or arguments between two parties over what is or is not acceptable. Universal standards are a fine ideal but often fail to provide for situations where certain types of equipments and other facilities are unavailable.

Another important but oft-neglected aspect of "standard' procedures concerns the question of terminology and definition. Most international standard bodies are in fact still arguing about terminology. It is a contentious issue amongst standard bodies and even among many leading experts. Consequently, in the scope of this chapter terminology will be adopted that is considered to be generally acceptable.

It is probably not generally appreciated that coal, due to its heterogeneous nature, is one of the most difficult bulk commodities from which to obtain a representative sample. Further to this complication, vessels can contain single and/or multiple consignments, each of which can have different sampling characteristics and require different sampling schemes and modes of analysis and preparation. Although most consignments of coal around the world would conform with a 50 × 0 mm size distribution, there may be particles ranging from 100 mm to less than 0.01 mm.

Sampling can be manual, mechanical or a combination of both. Manual sampling relies on the diligence of the individual to take increments in the correct sequence and to follow predetermined procedures. The main advantage of manual sampling is, that in daylight hours and well-lit areas, the sampler can also observe the coal and recognise major physical changes in moisture, size and evidence of contamination, which is much harder in fully integrated mechanical sampling systems. The major disadvantage is that it does rely on the individual and in some areas where it is difficult to take increments and strength is required, people may take biased samples. Manual sampling can best be accomplished in stationary situations where small tonnages are to be sampled. Manual sampling of stockpiles or from conveyors of greater than 300 t/h capacity are not recommended as the chances of taking biased samples are quite high and the dangers to the sampler cannot be justified.

Obviously, with large tonnages of coal being shipped vast distances around the world, an increasing number of loading and unloading ports are equipped with high capacity, high speed handling systems. In order to ensure that the vital sampling process does not impede (and therefore negate) the increased handling efficiency, mechanical sampling systems have been installed at such ports. They are designed to take increments from the moving streams in accordance with international standard bodies (ie ISO, BS and ASTM). These installations, after commissioning, should undergo bias tests in order to ensure that the levels of bias in total moisture, ash and calorific value are not significant. In real terms, one must ensure that the sample the mechanical system is producing is representative of the coal being loaded or discharged. Even after this test has been carried out it is not safe to assume that the system will remain unbiased. Regular maintenance checks together with occasional bias pair checks are strongly recommended. Before commencing a bias test an audit of the system should be undertaken to determine if the various parts of the mechanical system are actually doing what they are designed to do:

1. Determine primary increment mass.
2. Determine cutter aperture sizes.
3. Determine subsequent increment mass from secondary and tertiary cutters.
4. Ensure system can take standard number of cuts from primary.
5. Determine belt speeds of system.
6. Determine top size of crushed product, inspect hammers or rollers in crusher. If the crusher is a hammer type mill, ensure there is a breather pipe and that it is clear.

Many of the above checks should be regularly carried out to ensure the system is in good operational order.

A considerable amount of work is taking place in order to establish an agreed international standard for mechanical sampling systems. Much of the technical content of this standard is now in place, such as the organisation of the sampling procedures, major features of equipment, design and operation and other structural factors. The key questions involve the choice of what number of sampling units, number of increments in each unit and the frequency and mass of each increment. There are two methods used to adjust the frequency of increments in a mechanical system:

1. Time-based sampling.
2. Mass-based sampling.

Time-based sampling involves taking primary increments at regular time intervals while the coal is being loaded or discharged. It is suitable when loading rates are consistent and primary increment masses are also consistent.

Mass-based sampling involves adjusting the speed of the cutter in order to vary the primary increment mass according to the mass on the belt. This is achieved by connecting a weightometer reading to the primary increment system. This system is suitable when there is an inconsistent loading rate such as with large bucket wheel reclaimers. Many of the newer export ports are based on this particular system and mass-based systems are either in operation or recommended.

If the coal has been sampled mechanically or manually, it produces a series of separate increments and these will be combined to form sampling units. Mechanical systems have the advantage when each primary increment can be crushed and divided as it is taken.

The specification of some coals involves a stipulated minimum percentage over-size or a maximum under-size in a graded product. For such coals therefore size distribution is an important factor. Preferably a separate set of increments should be taken and combined to form either a single gross sample or a series of sampling units. The latter is preferable because then the mass of the material involved for each individual test is manageable and also such a procedure affords the possibility of performing a check on any variation within the sub-lots, which in turn provides a measure of the overall precision of the final result.

Manual and automatic systems are designed to produce representative samples that can be subjected to laboratory analysis. The tests normally performed are:

Proximate analysis:	Moisture in the analysis sample
	Ash
	Volatile matter
	Fixed carbon (by difference)
Ultimate analysis:	Carbon
	Hydrogen
	Nitrogen
	Sulphur
	Oxygen plus error (by difference)

In addition to these conventional tests there are specific tests to measure certain given quality criteria:

1. Ash analysis.
2. Calorific value.
3. Hardgrove grindability index.
4. Crucible swelling number.
5. Petrographic composition.
6. Reflectance measurements.
7. Audibert-Arnu dilatometer.
8. Gieseler plastometer.

From the above list will emerge a set of criteria by which the product can be evaluated for commercial purposes.

CHAPTER 13

Principles of sampling of coal

The generic term "coal" can cover a very wide spectrum of solid fuels. As the term "coal" in itself has such general connotations, it is usually traded under names that reflect the use to which it is going to be put. Consequently, it is the suitability of a given coal for eventual technical utilisation that will govern its classification. The most common designations one encounters are anthracite, steam-coal, coking coal and coke (gas coal).

This rather primitive classification can then be used to produce specifications based on usage, which will reflect the major characteristics of the coal in question. It therefore provides the purchaser with terms of reference in order to evaluate the quality of coal delivered.

Obviously, any evaluation process referring to quality and quantity of material delivered is going to involve some form of repeatable sampling process. Consequently, once the specifications of the coal in question have been defined, a sampling method must be devised which will select material representatively, and which can then be tested in order to reflect the key characteristics and enable them to be measured against an international specification. Before proceeding to a description of sampling methods available for coal, it is instructive to reiterate the nature of the mineral itself.

A coal consists of a mixture of three principal components:

1. Carbon (and carboniferous materials);
2. General mineral material;
3. Moisture.

The carboniferous materials, comprising carbon and carbon compounds are all combustible, whereas the second category, mineral materials are not combustible. The relevance of moisture is evident as it directly reduces the calorific value of the material.

This is a heterogeneous mix and therefore the above components are spread unevenly in the various size fractions of the coal. Generally speaking, the smaller particles will contain a higher proportion of mineral matter compared with the larger particles and will also have a much higher density than larger particles.

In classification terms, mineral material is usually referred to as ash which, whilst not literally correct, is a useful and practical label. This is due to the fact that the ash determination is the usual way of determining the mineral content and consequently is an integral part of chemical analysis of coal.

There are two characteristics that can affect the handling of coal during the various transit operations and for that reason have to be carefully monitored. The first concerns the particle size (granulometry), as well as the shape and density of the particles. For example, very fine material can be subject to excessive dusting losses due to wind and other adverse climatic conditions. The amount of particle breakdown is a measure of the brittleness of the material as well as the density variation that can give rise to segregation of cargoes, which in turn will cause sampling problems.

If the coal is very wet then a disproportionate amount of so-called free moisture may tend to settle in the lower layers of the cargo and by the same token the moisture content in the upper layers of the coal when stored outdoors can be affected by climatic conditions. This also poses problems when one has to obtain a representative sample from, for example, a stockpile.

GENERAL SAMPLING PROCEDURES

In view of the wide range of sampling philosophies available, it is more practical for a publication of this kind to consider the most frequently used international standard, which at present is ISO 1988. This applies to the manual sampling of hard coal. For mechanical sampling the ISO TC27/SC4 will be used as the reference point.

TIME AND PLACE

The two most critical criteria that apply to the establishment of a sampling procedure at a given location are time and place.

As a general principle it is common sense that sampling should take place as near as possible to the point of weighing. This can either mean that sampling has to be carried out close to a weigh-scale, or if the weight is determined by draft survey, sampling should take place either during the whole of the loading (or discharge) operation.

An essential condition applied to the sampling process is that, if possible, the whole of the cargo to be sampled should be available to the sampler. This means that all parts are equally accessible to the sampler (or the mechanical equipment used) and therefore all parts of the cargo have an equal chance of being sampled.

Ideally the following locations ("places") are listed in descending order of preference:

1. Stationary conveyor belt;
2. Falling stream (preferably at conveyor belt transfer point);
3. Manually from barges, lorries or wagons;

4. From ocean going vessels;
5. Stockpile.

The latter—stockpile sampling—is not recommended and should only be considered as a last resort.

There are other factors that apply to the choice of the sampling location, particularly accessibility and safety. There must be adequate space available for handling the sampling equipment and logistically there should be facility for the collection and storage of sample containers with little or no risk of loss, contamination or degradation of sampled coal.

With regard to time, the decision has to be taken regarding the intervals at which sample increments are taken from the cargo. There are three basic methods of deciding the time interval:

1. A systematic sampling scheme whereby the increments are spaced evenly either on a time or weight basis.
2. Random sampling whereby increments are taken randomly during loading (or discharge operations).
3. So-called stratified random sampling. In this case the sampling operation is divided on either a time or weight basis into an equal number of strata in the cargo. Then, increments are taken at random from each of the strata.

Obviously, systematic sampling is the most preferable procedure and the one most frequently adopted. The only danger to such a procedure is the existence of any periodic variation in quality in the cargo which, if it should be in phase with the chosen sampling frequency, could lead to a bias either in moisture or size or any of the analytical characteristics. Such a factor is normally avoided by the experience of the sampling inspectors involved.

NUMBER OF SAMPLE INCREMENTS

A mass of mathematical calculation and theory has been devoted to the question of the number of increments required to achieve the desired precision of sampling. It has been said that sampling and statistics must be inseparable because the interpretation of accuracy and precision data requires a basic knowledge and understanding of statistics. However, the elaborate and often elegant mathematical models of those such as Pierre Gy will not be dealt with in this text. Any reader wishing to know more about the statistical background of sampling as well as understanding the principles of applied statistics including items such as propagation of variances, time series analysis, correlation-regression analysis and the analysis of variance, should be referred to Jan Merks's fine book *Sampling and Weighing of Bulk Solids.*[1] For this text only the practical applications of such statistical models will be considered, together with the way in which the results of such studies have contributed to the establishment of the international specifications presently in use.

The number of increments required in order to obtain a given desired precision is a function of the variability of the quality in the coal, irrespective of the weight of the lot or sampling unit in question. The measure of such variability is, however, the critical component in the evaluation. Often, this primary increment variance has to be obtained and often refined on the basis of experience. Data obtained from a given mine production together with test results from sampling relatively small sampling lots can provide a primary increment variance specific to a given grade of coal. However, this idealised approach cannot be considered sacrosanct. During transportation of very large masses of coal, segregation will inevitably occur and changes in quality may result.

Similarly, if coal is stockpiled for long periods in varying climatic conditions, similar changes of quality may occur.

However, taking the ISO 1988 as an idealised basis, the number of increments required for the precision requirements for moisture and ash contents on reference standards are as set out below in Table 82.

In each case we will consider the standard of precision achieved for the ash and moisture with less than and more than the 20% threshold value for both ash and moisture. ISO 1988 bases the number of increments required on the reference standards of precision for moisture and ash contents shown in Table 82.

Table 82 Reference standards of precision for moisture and ash content

Characteristic	Type of Coal	Standard of Precision
Ash	<20%	± one tenth of true ash
	>20%	± 2% absolute
Moisture	<20%	± one tenth of true moisture
	>20%	± 2% absolute

Considering consignments of less than 1,000 tonnes, the number of increments recommended in order to achieve the desired precision for ash determination are shown in Table 83.

Similarly, if sampling for total moisture only, the number of increments required, relating to the type of the coal are given in Table 84.

1. *Sampling and Weighing of Bulk Solids* by J.W. Merks (1st ed., 1985, Trans Tech Publications).

Table 83 Sampling place

Type of Coal	Stopped Belts and Falling Streams	Wagons, Trucks and Barges	Sea Going Ships	Stockpile
Cleaned	16	24	32	32
Uncleaned	32	48	64	64

Table 84 Number of increments required when sampling for total moisture only

Type of Coal	Number of increments
Unwashed or Dry Coal	16
Washed Graded Coal	16
Washed Smalls	32

For larger consignments (ie in excess of 1,000 tonnes) there are two principal options:

1. The consignment be divided into sampling units of 1,000 tonnes or less, from each of which a separate sample is taken with a specific number of increments.

2. Alternatively, the consignment may be sampled as one lot or the consignment be divided into a number of sampling units each of more than 1,000 tonnes. In either case the initial number of increments is multiplied by the following empirical factor:

$$\sqrt{\frac{\text{mass of unit (in tonnes)}}{1000}}$$

Taking a theoretical sample, the minimum number of increments required for ash and general analysis can be shown in Table 85.

Table 85 Minimum number of increments required for ash and general analysis

Size of sampling unit	Stopped Belts and Falling Streams		Wagons, Trucks and Barges		Sea-going Ships and Stockpiles	
(tonnes)	Cleaned	Uncleaned	Cleaned	Uncleaned	Cleaned	Uncleaned
2,000	23	46	34	68	45	90
5,000	36	72	55	110	70	140
10,000	50	100	75	150	100	200
20,000	70	140	110	220	145	290
50,000	115	230	170	340	225	450
100,000	160	320	240	480	320	640

The minimum number of increments required for total moisture determination is identical to that specified for ash from stopped belt or falling streams.

The reference point for size analysis is ISO 1953. According to this standard, the minimum number of increments required for combining into a single test sample is 40. If however, the nominal top size is large (ie in excess of 80 mm) it is usually recommended that six replicate test samples are taken. Each test sample should comprise at least five increments in order to provide a reliable check on the precision of the sampling process.

When considering the minimum mass of increments, ISO 1988 specifies an empirical formula for coals up to 150 mm nominal top size. In this formula the minimum mass of increment (P kg) is determined thus:

$$P \text{ (kg)} = 0.06 \text{ D (mm)}$$

where D is the nominal top size (in mm). It is stipulated that P should not be less than 0.5 kg.

For mechanical sampling processes, the minimum width of cut from either a stopped belt or a falling stream should be at least three times the nominal top size of the coal.

For manual sampling, the minimum width of sampling scoops, augers or probes should also be at least three times the nominal top size of the coal. In practical terms, this means the width of both the mechanical cutter or the manual sampling equipment should never be less than 30 mm in order to avoid clogging. It should be noted that as a general rule, manual sampling of coal with a nominal top size in excess of 80 mm should only be done when the material is stationary.

During the preparation of the mechanical sampling document, questions have been raised on the validity of the empirical formula used for calculating the number of increments and minimum mass of increment. Not only does this apply to the primary increment, which in mechanical sampling is usually exceeded, but also to sample preparation when large masses of gross samples or partial samples need to be reduced and divided to acceptable low masses for laboratory samples or test samples.

DEVISING A SAMPLING SCHEME

For the purposes of this discussion, we need only consider cargo sizes in excess of 5,000 tonnes as it is generally accepted that anything less than such a tonnage can be treated as a single sampling lot.

Shipments in excess of 5,000 tonnes are usually divided into a convenient number of sampling units. On this basis, a formula for estimating the minimum number of sampling increments in a given lot is:

$$m = \sqrt{\frac{L}{5000}} \text{(taken to the nearest integer)}$$

where L is the mass of the lot in tonnes.

This is only a minimum number and can be increased on the initiative of sampling inspectors according to either convenient weight or convenient time factor or due to any fears over issues such as the periodic variation in quality mentioned above.

Again, based on experience, the number of increments per sampling unit should not be less than ten. The sampling units are samples representing various tonnage levels within the lot, ie 30,000 tonne lot with ten sampling units of 3,000 tonnes. This ensures a good precision is obtained and, coupled with the flexibility of being able to adjust the number of sampling units and the number of increments per sampling unit, any bias resulting from the storage, handling or sample preparation of very large bulk samples can be minimised.

A very important component of the sampling scheme is the interval of sampling. This can be calculated in one of two ways depending upon whether the emphasis should be placed on time or mass of material. The formulae are empirical but have been tried and tested over many years.

If the emphasis is to be on time, then the time-based sampling interval in minutes is calculated from:

$$\frac{\text{Interval}}{\text{(in min)}} = \frac{\text{Sampling Unit (tonnes)} \times 60}{\text{Av. Flow Rate (t.p.h.)} \times \text{Min. no. of increments}}$$

If the emphasis is on mass, then the mass-based sampling interval is:

$$\frac{\text{Interval}}{\text{(in tonnes)}} = \frac{\text{Mass of Sampling unit (tonnes)}}{\text{Min. number of increments}}$$

It should be stressed that in devising a sampling scheme, and particularly when considering the optimum sampling interval (whether it be based on time or mass), due emphasis must be placed on a possible variation in quality of the material being sampled.

Once again, it is instructive to consider the two general processes involved in drawing samples, manual and mechanical.

With manual sampling, the mass of each individual increment is usually consistent. There is no significant improvement in precision in drawing samples that are larger than the minimum required. Obviously one should never take fewer increments of larger mass.

In the case of mechanical sampling, in practice the minimum cutter width and the maximum cutting speed when used in high capacity belt conveyors provide primary increments with masses well in excess of the minimum required. However, with most systems, each primary increment can be of the order of 200–600 kg. In such cases, the bulk sample is not merely passed to the laboratory, but has to be divided, sometimes with reduction, and prepared to give a smaller mass for the laboratory to analyse. Obviously, large lots must be divided into sampling units in order to ease any problems that might arise from collection and storage of prepared samples. In addition, it will help to minimise the possibility of moisture loss during handling and storage, as well as improve the precision of the final analysis. It is usual to extract a separate sample for the determination of total moisture from each sampling unit.

As a general point, the decision taken on the number of sampling increments must be weighed against the possible improvement in precision on the one hand and economic factors on the other. Clearly, an over elaborate process is likely to involve longer time and consequently delays in discharging or loading the vessel and possibly involving more labour. All these economic factors have to be taken into consideration.

Equally, whilst empirically an increase in the number of increments should improve the precision, if such improvements are only marginal, then the extra cost involved may actually be counter-productive.

CHAPTER 14

Manual sampling

As stated previously, the selection of an appropriate sampling scheme and the equipment and location of sampling operations must all be aimed at minimising the possibility of any bias in the drawing of samples.

As this is not a comprehensive review of sampling *per se*, only the principal methods that are available for manual sampling will be considered. Drawing a sample from a stopped belt is the only accepted way of ensuring that all component parts of the material have a reliable chance of being collected. In this way, the chances of obtaining an unbiased sample are extremely high. Consequently, this method can be reliably regarded as the reference against which all other procedures will be judged.

SAMPLING FROM A STATIONARY BELT

This procedure obviously involves drawing a section of the material from a conveyor belt that has been stopped at a prescribed time/weight interval. Obviously, if the flow of material onto the belt is regular, then the samples can be extracted at equal time intervals. If the flow is irregular however, the sampling interval will probably be phased according to the weight loaded onto the belt.

Samples are normally withdrawn by means of a sampling frame, the width of which should be at least three times the nominal top size of the coal. Once again, safety is an important criterion which is why sampling from a stopped belt is preferable. Also, it is the most secure procedure in that the sampler can ensure that all material at the particular sampling point is taken as the incremental sample.

SAMPLING FROM A FALLING STREAM

It has to be recognised that it is not always operationally and/or economically feasible to stop a conveyor belt in order for samples to be drawn. Therefore, the next logical position from which to obtain a reliable cross-section of material flowing along the belt would be at a transfer point from one belt to the next. At such points the coal will be falling freely.

In such situations for manual sampling, a simple sampling ladle can be applied, which is moved at an even speed, once across the stream of falling coal. The critical factor here is the speed of the flow of the coal. If this is too fast, then there is a risk that the sampling ladle will be filled before it has reached the full width of the stream. The only remedy to such a situation is to increase the size of the ladle. Normally, as with the sampling frame mentioned above, the width of the ladle should be approximately three times the nominal top size. Simply increasing the size of the ladle is not the complete answer as one may reach a situation where the ladle size is too big to traverse the full stream manually. This problem can be overcome by the application of mechanical assistance where the ladles can be suspended on rails and pushed or pulled through the stream. However, this should never be attempted on conveyor belts with a speed greater than 300 tonnes per hour.

As well as providing a risk that the ladle will be filled before it has crossed the stream, a fast moving belt is also of possible danger to the sampler.

It does not take a lot of imagination to realise that sampling from conveyor belts is easily mechanised. The whole area of mechanical sampling will be dealt with later.

SAMPLING FROM RAILWAY WAGONS OR ROAD TRUCKS

Again, applying the principle that it is always more satisfactory to sample moving material, if one has to draw samples from either wagon or truck, it is better that this be performed either during loading or unloading. If, however, samples have to be taken from a full wagon/truck, then great care must be taken to ensure that all layers of material in transportation are sampled. This is particularly critical for the total moisture content as it is often segregated either to the bottom or to one of the sides.

Samples can either be drawn using a tube auger or drill auger depending upon the size and the moisture content of the coal in question. Obviously, it is essential that all layers are represented in the sample, so that whatever implement is used, it must go right through the material in the wagon/truck. Particular attention has to be paid to any large lumps present, which might segregate along the sides or the end of the wagon/truck.

One of the greatest dangers to this type of situation involves what is called "car top sampling". In such a situation, for reasons of "economy" (of time and effort), only the top material on each rail wagon or truck is taken. Clearly, there is a risk in such situations that if the sampling is taking place preshipment, and that by either accident or design, the best quality coal goes on top of each piece of transport, a biased sample towards higher quality is taken for evaluation purposes.

SAMPLING FROM A CRANE GRAB

Obviously, this procedure is only applied when it is either uneconomic or otherwise impossible to sample from conveyor belts. However, there is a greater danger of systematic errors occurring from sampling from grabs as opposed to belts.

For purposes of sampling, the grab must be stationary and preferably resting on the ground. The outer layer of coal is pushed aside before sampling occurs either by means of a spear or a special high sided shovel. The reason for removing the top 20cm or so of outer layer is that this material will normally comprise coal which has come down from the top of the grab.

SAMPLING FROM A SHIP

This procedure is hardly ever used, both on logistical and safety grounds. It is most unsatisfactory from the point of view of obtaining a sample truly representative of the cargo. Once again, the two criteria of size and moisture represent the greatest risks. Large lumps tend to gather along the sides of the hold and the lower layers of the cargo are generally more moist than upper layers due to drainage. Therefore, there is a very grave risk of major errors occurring in the percentage of moisture and ash reported in the final quality analysis.

Rarely, if ever, these days will sampling be performed from deep holds in ocean going vessels. However, there may be occasions when sampling is required from material that is in barges or other flat-bottomed vessels. In such cases, when the cargo does not exceed more than 4 m in depth, the material may be sampled using an auger. Once again, this is not the most satisfactory way of sampling the cargo, but it may be the only one feasible in the circumstances (especially in relation to time difficulties).

SAMPLING FROM STOCKPILE

This is extremely unsatisfactory as, clearly, only a limited proportion of the coal is accessible for sampling. This problem is exacerbated when the stockpile has been in existence for a long period of time. Such stockpiles are normally extremely heterogeneous because they probably comprise several different grades of coal which have been accumulated over a period of time. The reason why stockpile sampling may be invoked in such situations, would be where a middleman has accumulated material and then decides to sell it on at a later stage and is therefore not the end consumer. The potential buyer in such a situation is therefore severely at risk and the cost involved in conducting a proper sampling (ie which would involve moving the entire stockpile) might be prohibitive.

Where such commercial/economic factors apply and stockpile sampling is the only method that will be contemplated by the potential customer, a tube auger will normally be used. The coal must be sufficiently wet to hold within the tube. Samples would normally be taken from the top, middle and bottom of the stockpile, usually in the rate 3:5:7.

If the coal is insufficiently moist to enable an auger to be used, material will have to be dug out from the pile at the prescribed locations and samplers will have to use their own experience to judge whether, in fact, they are drawing representative samples.

It cannot be stressed too strongly that such a procedure is extremely risky and is very unlikely to produce a truly representative sample.

If a conveyor belt forms part of the coal transfer system, sampling can be carried out by the stopped-belt method. In view of the fact that the stockpile is most probably heterogeneous, the number of increments taken should be a minimum of 64 for stockpiles up to 1,000 tonnes. Such a number is necessary even though the samples are being taken from a conveyor belt in order to overcome bias due to the heterogeneous nature of the material.

Auger sampling of stockpiles greater than 20,000 tonnes would give results likely to be biased. In such cases the best method would be to take increments from the cut-face of the stockpile as it is being loaded or shifted.

CHAPTER 15

Mechanical sampling

Whole books have been written on this subject and there is, of course, a wide range of different devices in existence. Therefore, for the purposes of this particular treatise, only the general features of such systems will be considered.

First and foremost, the sampling installation must have sufficient capacity to retain completely or entirely pass any increment without spillage. There are various other additional criteria that are essential to an efficient installation. It must remain self-clearing if the sample has eventually to be used for size determination; it must additionally minimise any fluctuation in moisture content; it must minimise the possibility of loss of fine particle material; and finally, the equipment must be capable of handling a wide range of types of coal likely to be passed through the location in question.

Some mechanical sampling systems take only a particular section of a moving stream and this is not considered satisfactory. As a consequence, we will limit ourselves to processes that take a complete cross-section of the falling stream in one movement of a primary cutter system.

With this in view, clearly, the primary sampling equipment must be situated at a point where the whole stream is accessible. The primary sampler should be situated as closely as possible to either the loading or the unloading point of either the conveyor belt or other moving system.

There is a widespread belief that as with modern instrumentation in chemical laboratories, the installation of any kind of automatic electronic equipment is foolproof. This is patently not the case, and therefore any mechanical sampling system must provide the facility for checking against bias. The system must be designed in order that replicate samples can be drawn at the primary cutting stage so that they can be checked for precision.

Obviously, safeguards must be implemented to prevent contamination and for this reason the installation should be so constructed as to avoid spillage of samples or the build-up of coal at any point. This causes great difficulties in keeping the system clean, which means that samples from previous cargoes could be introduced into the test samples.

The most important component of a mechanical sampling system is the primary sampler. Most of these are designed to collect increments from falling streams at transfer points in a conveyor belt system. The two critical criteria in designing a primary sampler are the width of the collection shoot and the speed of the cutter entering the stream. As mentioned above, the primary cutter must be designed in order to take a complete cross-section of the falling stream of coal. The leading and trailing edges of the primary cutter usually follow a path describing an arc normal to the path of the falling stream.

Usually, the primary cutter will travel through the falling stream at a uniform velocity designed to provide an equal exposure time to any point in the stream. However, many new sampling systems are mass-based and as such the primary sampler velocity varies according to the tonnage on the belts.

Conventionally, the width of the sampling aperture is a minimum of three times the nominal top size of the coal being sampled. The critical criterion is that the procedure must be shown to be free from bias. In order to avoid the risk of clogging, the minimum aperture of any primary cutter should be 30 mm.

The question of the speed of the primary cutter often requires a degree of trial and error at the design stage. It must also be designed in the context of the type of coal for which it is most likely to be used. If the cutter velocity is too fast, it will cause deflection of larger particles and therefore introduce an immediate bias in the sampling.

Inevitably, there is a whole wealth of sophisticated mathematics which has devised formulae linking the flow rate of the material on the conveyor belt, the size of the cutting aperture and the speed of the aperture passing through the falling stream to the mass of the primary increment.

It cannot be stressed too strongly that regular tests must be performed on such installations in order to prevent sample bias. Generally speaking, the reference point for checking bias in such systems is to use stopped belt sampling as described earlier.

For installations that are in almost constant use, regular inspection and maintenance is carried, particularly looking for building up of coal of any factors that could contribute to restriction of flow of material through the system. Once again, from a point of view of avoiding contamination, a mechanical system should be flushed comprehensively with any new grade of coal before such material is sampled on that system.

Several times in this text the terms "precision" and "bias" have been used quite freely and it is therefore important that these terms are understood as in the minds of some people they can appear synonymous with the word "accuracy".

Precision really describes how closely repeated measurements agree among themselves. Consequently, if all the repeated measurements are systematically in error, then such measures will show a high degree of precision but would be both biased and inaccurate.

Bias reflects the difference between a theoretically correct value and the statistical average of a number of measurements. Such a difference is equal to the mean of the error distribution. It therefore reflects the extent to which the measurements systematically vary from a theoretical norm. Accuracy is a measure of the absence of bias with results that show good internal agreement (precision).

The danger of applying statistical formulae to the measurement and checking of the precision of the performance of primary samplers in mechanical systems, is that they make certain assumptions which are not always valid. For example, there is an assumption that key quality criteria of the coal vary randomly throughout the given cargo. There is often also the assumption that, providing sufficient readings are taken, these will follow a normal statistical distribution. Neither of these assumptions are valid. For example, a stream of coal can show long-term variations which will be superimposed upon random short-term variations, such as the systematic change of quality in a seam of coal. These, in turn, will be superimposed on changes in quality due to degradation of coal during handling and/or variations in moisture. In such cases, estimating the precision of a primary sample based on increment variance and preparation and testing variance will produce a precision figure that is numerically higher than the precision actually obtained. Therefore, it is essential that such systems are evaluated and checked using replicate sampling procedures. These statements can be confirmed and demonstrated using statistical formulae which are outside the scope of this treatise.

TESTING FOR BIAS

Clearly, before bias tests can be performed, the sampling system must be thoroughly examined and all variances known or at least suspected. Such an examination is long-winded and therefore costly and procedures have to be followed rigorously.

In essence, a bias test programme seeks to determine whether a given mechanical sampling system collects samples that are statistically consistent with reference samples taken from a stationary conveyor belt. The tests usually applied in order to assess such consistency are: total moisture, ash, sulphur and calorific value. The statistical significance of the analytical differences is determined by using the familiar student's "t" test.

CHAPTER 16

Analysis

SAMPLE PREPARATION

Once a reliable primary sample is obtained by any of the procedures described in Chapters 14 and 15, the bulk sample is taken for final preparation, the object of which is to produce material in a form that can be subject to analytical tests in the laboratory.

The procedures involved for sample preparation have to relate to the characteristics of the coal in question. Other key factors are the total weight of gross (bulk) sample, the nominal top size and the moisture content. If the sample is visibly wet, it is usual to subject it to a preliminary air drying process. Moisture loss during this process is determined and added to the overall total moisture content. The final product for most laboratory tests should normally be material (reduced) to a size of minus 0.212 mm with a minimum of fines. The question of reduction is of importance as most laboratories require the order of 60 grammes (minimum) of material for analysis. Ring grinders or ball mills should be avoided with coal; small hammer mills are much preferred for the

Plate 10 Hammer mill.

reduction process. (a typical hammer mill is shown in Plate 10). It is always better to avoid grinding.

In addition to chemical tests there may be special tests required such as Hardgrove grindability index, abrasion index, the Gieseler plastometer test, petrographic analysis and dilatometer test. It may be necessary to take extra material for some of these tests or take the unused material from the various reduction stages in the sample preparation process.

The key stages in sample preparation (often following air drying) are:

1. *Size reduction* — by crushing and milling.
2. *Mixing* — to achieve homogeneity.
3. *Division* — to decrease the overall weight of sample being handled

The purpose of air drying is to ensure that all sample material will pass through the reduction and sample division equipment freely without either loss, contamination or serious change in moisture content.

Usually air drying involves the material being spread out to a depth not exceeding twice the nominal top size, in trays which are then placed in ovens where heated air is passed over the coal at maximum temperatures controlled normally around 40°C. If calorific value or various carbonisation tests are to be done, then the temperature should not exceed 30°C. However, there are various options. For example, the time can be restricted as well as the temperature and therefore combinations that can be applied are:

- 40°C for three hours.
- 30°C for six hours.
- 25°C for 24 hours.

Preparation normally consists of crushing and dividing to produce a representative sample of coal suitable for analysis. Crusher types are hammer mills, roll crushers, plate mills, ball mills and ring grinders. For the preparation of coal, impact type mills are recommended but not grinding type mills such as ring grinders and ball mills.

The main reason for this is that grinding type mills tend to produce heat, which may affect the properties of the coal you wish to test. Calorific value and carbonisation tests may be adversely affected by the use of grinding type mills.

One common fallacy is that crushers not only reduce top size, but also mix the sample. Hammer mills in particular tend to segregate the sample due to the fact that the hardness of the coal varies from size to size. Generally, the finer coal is softer and will progress through the mill first and the larger particles, which are generally harder, will be the last to progress through the mill. Thus in your sample container under the mill you will have a segregated sample which will have to be mixed adequately before proceeding with any analysis.

Many tests are carried out on coal and these require various different modes of preparation. The top size of the coal sample can vary from between 10 mm down to − 212 micron. To prepare the final analysis sample for the laboratory to − 212 micron top size, the most commonly used mill in Australia is the Raymond mill (see Plate 11). It is a small hammer type mill with a screen to control size distribution of the final sample. It is very important to determine the nature of the coals being handled. Soft coals may not require the level of crushing to achieve the required size distribution for analysis compared with very hard coals. The Raymond mill has provision for eight hammers but can be operated with four. Regular checking of the mill screen and the size distribution being attained is essential.

Plate 11 Raymond mill

The reason for this is that the screens can become blinded and/or holed. This will either cause an increase in the + 212 micron or an increase in the fines. With an increase in + 212 micron it is common to see poor duplication in basic analysis such as ash and volatile matter—in technical terms, a lowering of the analysis precision. If the screen is blinded and the sample has too many fines, physical tests such a crucible swelling number may be affected. The important point to remember is that crushers must be adjusted to achieve the size distribution required for analysis and these adjustments checked regularly. Crushers segregate coal samples and mixing should always follow crushing. Cleaning mills between samples is essential to avoid contamination.

It is also very important that the mills involved should not be operated at such a speed where the material can overheat. As is standard with such a process, the equipment must be easy to clean and must be inspected and cleaned at regular intervals and always flushed out thoroughly with a sample of the material to be prepared.

The point raised regarding minimising the generation of heat, means that compression mills (eg jaw crushers or roll crushers) or impact equipment such as hammer mills are preferable to ring mills or plate mills which tend to generate more heat.

Similarly, hammer mills should not be used where excessive fines are to be avoided for example, for samples required for Hardgrove or petrographic tests.

A critical stage in the sample preparation process is the mixing of the materials. There are several processes available for the manual division of the quality sample. The most common are listed below:

1. Increment division.
2. Rotary sample division.
3. Riffling.
4. Mixing and splitting.
5. Coning and quartering.
6. Fractional shovelling.

Some of the above are traditional techniques employed with other materials which have been transferred to coal. Coning and quartering is an example of a dividing technique widely used in the sampling of other materials. Unfortunately, in practical terms one tends continually to segregate the sample due to larger pieces rolling down. Many of the more recent up-dates of coal sampling standards have deleted this method almost entirely.

Manual increment division and/or fractional shovelling offer the best technique for manual division. Riffles can be used but there are always difficulties in sample distribution prior to division. Riffles have trouble dividing damp coals due to sticking and blinding of apertures. Many modern preparation areas use rotary sample dividers as a means of division of coal (a typical divider is shown in Plate 12). These have many advantages as follows:

1. The division is a constant one and not operator dependent.
2. The division method is easily checked for errors using duplicate samples.
3. Operated correctly, it provides almost constant weight division.
4. It can divide damp coals without sticking or blinding.
5. It is much faster than conventional manual techniques.

In conclusion, for preparation of coal samples, hammer and/or roll crushers should be used for crushing and rotary sample dividers for division.

Plate 12 Rotary divider.

TOTAL MOISTURE CONTENT

As with the determination of the moisture content of any bulk material, the main difficulty is to reduce the possibility of changes of moisture during handling, storage and transportation stages.

Preparation of samples for total moisture

The principal difficulty in the determination of total moisture content is the avoidance of any reduction in the total moisture by handling, storage or transporting the samples. This is a very common cause of discrepancies in moisture results.

There are three main methods for the determination of total moisture:

1. Distillation with toluene in a calibrated container.
2. Drying in an oven in an atmosphere of nitrogen.
3. Drying in an oven with frequent and adequate changes of air.

It is probably true to say that most independent laboratories prefer to use the third option for the determination of total moisture. The second method is very widely used for the determination of moisture in the sample for analysis.

The first two methods apply to all hard coals and for this a minimum sample of 300 g is required of coal less than 2.8 mm in size. The third method applies especially to coals that are not susceptible to oxidation and require larger samples. For example:

> 1.0 kg for top size 20 mm
> 0.6 kg for top size 10 mm

However, a lot depends on the inspector's own assessment of the appearance of the coal. For example, if it is visibly dry and can be processed quickly without any undue loss of moisture content, a one-stage determination is possible. On the other hand, if the coal is visibly wet, it is advisable to apply a two-stage method (unless a closed mill is available), which is capable of crushing the sample without loss of moisture content.

Air drying can be necessary at any stage in sample preparation depending upon the visible moisture of the coal bulk sample. Obviously, if air drying is applied, then the moisture loss must be carefully calculated and taken into account when calculating the final total moisture content.

Preparation of samples for chemical analysis

As mentioned above, the first step in the preparation is air drying and that should ensure that the whole of the bulk sample will pass freely through any milling or subsequent separation and dividing stages.

As with all samples for analysis, the materials for testing must be always packed very carefully to ensure no damage or contamination arises. Also, samples must be clearly and unambiguously marked with all references including locations and identification of material.

Sizing

The portion of the bulk sample set aside for granulometry tests needs only to be air dried, and this process is only necessary for coal samples that are too wet for accurate sieving and division.

The procedures used for sieving are obviously dependent upon the type of coal, the size range in the cargo and the specification against which the coal is to be tested. It is generally accepted that hand sieving is a good reference method. Mechanical sieving is acceptable for testing below the 4 mm size level, providing an end point test has been done previously. The danger of mechanical sieving for coarser sizes (above 4 mm) is that it is prone to bias because of breakages during sieving.

Many coals with particle size greater than 4 mm are sized using the hand-placing sieving technique. This involves the coal particle being turned around and passing through the aperture when upended. For example, a piece of coal 100 mm long and 45 mm in circumference will fit through a 50 mm aperture, and is therefore classified as minus 50 mm even though one of its dimensions is 100 mm.

For coals with a large top size and a wide range of sizes within the cargo, it is generally convenient to separate the test sample at around 40 mm in size. The procedure then is very straightforward; the oversize material is tested on a series of sieves starting with the larger size, and at each stage the fraction is collected into a tared receiving dish and weighed in order to obtain the weight of the individual fractions. The undersized (less than 40 mm) material is also weighed and if necessary, reduced in order to avoid excessive overloading of the finer mesh sieves. At this stage however, judgement must be brought into play. The inspector must examine the material and if in his view there is an excessive amount of fine dusty material, it is better to sieve out the main bulk of the fines fraction before carrying out the full sieve test. An illustration of sizing apparatus is shown in Plate 13.

There are various conventions that apply on a routine basis to the establishment of a size analysis. Usually, a continuous range of sieves are used and the sizes are chosen so that not more than 5% passes through the smallest size sieve. Generally speaking, it is not sensible to have more than 25% of the material being retained between any pair of sieves. There is also a rigourous procedure which involves shaking the sieve by hand, lightly tapping the material through the sieve, followed by rotation and inclination of the sieve usually repeated three times in order to have consistent procedure at each sieve size. This reduces the possibility of bias or distortion in the granulometry results.

Plate 13 Sizing apparatus.

In addition, there must be consistency over the shape of the hole used for the set of sieves. In other words, there must be a continuous range of one shape either square hole or round hole for consistency purposes.

As mentioned above, it is possible to use mechanical sieving for coarse sizes, although there is a risk of bias. Therefore, any mechanical system must be checked against hand sieving to ensure that no bias exists in the system.

Other key points when evaluating a mechanical sieving system are:

1. Sieving must continue until the same end point is obtained as that achieved by hand sieving.

2. The undersides of the sieves should be brushed regularly during the test (probably every two or three minutes).

3. The sieving action must be continuous and not so rigorous that the materials pass through the sieves due to degradation.

Usually, size analysis is reported by means of the weight of each fraction retained at the different sieve sizes and reported against the initial weight. If at the end of the analysis there is a deviation of less than 1 % of the total, the adjustment is usually made on the smallest size fraction. However, any deviation greater than 1 % means that the tests must be repeated.

ANALYSIS PROCEDURES

All of the elaborate procedures with regard to sampling, sample preparation and the intermediate handling stages are designed to produce material in a suitable form for chemists to perform the final quality evaluation process.

A lot of emphasis has been placed on sampling and preparation mainly because, in theory, most of the errors that can occur happen in these areas. Final analysis, however, also has its problems and should be considered as a major part of the system. The potential for bias is varied in the laboratory. Equipment has to be regularly calibrated and in good repair. Balances have to be checked and operators tested to ensure correct usage. As an example of precision and bias in the laboratory, correct use of the balance is important. If the balance is used incorrectly, but consistently incorrectly, a precise result but a biased result may be achieved. Also, if the balance is reading high or low, a precise result, ie duplicate analysis in the same test, but nonetheless biased result will be the outcome. Careful training of personnel in the laboratory is essential to ensure that differences in analytical technique are minimised. Regular checking of personnel with the use of known standards is a recommended practice and may highlight areas of bias. External round-robin exercises with other laboratories are also good methods of investigation to ensure methods employed are on the right track.

An important facet of the analysis of coal is the integrity of the sample. Reserves of samples at greater than 10 mm should be retained for check analysis and reference purposes. Samples at -212 micron top size, if susceptible to oxidation, may only be relevant for several days with regard to many tests such as calorific value. Check analysis should then be undertaken from a freshly crushed sub-sample from the reserve. Reserve samples should be retained in a freezer to maintain the original state of the sample as much as possible. Generally, the finer the sample is crushed, the larger the surface area and the more susceptible to oxidation the sample becomes.

Many of the tests are empirical and require strict adherence to the standard conditions. Volatile matter specifies a strict temperature of 900 $+/-$ 5°C for ISO and BS Standards. These temperature zones must be very accurately determined with a potentiometer, as slight variations in temperature cause differences in the level of volatile matter obtained. Likewise, crucible swelling number relies on the temperature of the burners being consistent. Many laboratories find a need to calibrate burners every time a bottle of gas is changed, as there can be a slight difference in Calorific Value of gas between bottles. Mains gas is not recommended for this test as the calorific value of mains gas may not be consistent enough to maintain the correct temperature on all the CSN burners.

Thus it can be seen that in the coal laboratory a continuing effort must be made to ensure equipment remains in calibration, balances are operating well, staff is working consistently and the quality of results remains high. Diligence and resilience at times are required as problems can be quite difficult to locate and solve. It can take years of experience to ensure the well running of a coal laboratory, and even then problems will arise from time to time that will require a lot of work and research to solve.

In conclusion, the very nature of coal provides a challenge to the coal analyst. Sampling and preparation of coal is tricky, onerous and at times a very dusty and dirty occupation. The analyses carried out in the laboratory provide a host of problem areas. This can provide quite a challenge to the coal sampler and analyst and requires a great degree of diligence and resilience.

PROXIMATE ANALYSIS

On arrival in the laboratory, the first stage is to ensure that all identification data on the package is totally compatible with the paper-work describing the origination of the cargo. It is particularly important to ensure that there is no anomaly or discrepancy over the total moisture content. On this point, it is important to know whether or not air drying has been performed at any stage. Most laboratories and major terminals analyse for total moisture directly from the -10 mm coal sample produced by a mechanical sampler.

The usual tests performed, particularly at pre-shipment, in order to verify that the coal is of suitable grade to be shipped, are termed the proximate analysis. This is merely a contraction of the word "approximate" as originally that was literally the case.

The components of the proximate analysis are:

1. Moisture in the analysis sample.
2. Ash.
3. Volatile material.
4. Fixed carbon (by difference).

Total moisture

There are two components of the total moisture content of coal. One is *hygroscopic* moisture, which does not influence the transportation in any way, and the second is *surface* moisture which could cause problems particularly with small coal particles. In some environmentally sensitive locations coal, that is subject to dusting, is sprayed in order to prevent excessive dusting and the water added is usually carefully metered. The studies carried out by the National Coal Board in the UK have shown that coal with a size distribution showing 20% of the material finer than 0.5 mm must be categorised as "difficult to handle" if the surface moisture is between 10 and 14% (this is not uncommon).

Another factor relevant to the surface moisture content of the coal is apparent in extremely cold weather because there is a danger that the coal will freeze, making handling extremely difficult.

The hygroscopic moisture is often referred to as the "inherent" moisture especially in a commercial context. The importance of this figure is that when the coal is burned, the ignition is delayed and flames are lengthened when the moisture content is increased. Equally, such coal will usually have a lower heating value.

There are three procedures normally used for moisture content:

1. Distillation using Dean and Starke equipment with toluene as the liquid medium. When coal is boiled with toluene, the moisture is carried over to the graduated receiver and calculated from a volume of water and mass of coal.

2. The material can be dried in an oven at 105°C in a current of nitrogen. The moisture content can simply be calculated from the loss of weight.

3. Also, the material can be dried in an oven at 105°C in air and the moisture calculated from the loss of weight. This method can only be used for coals that are not subject to oxidation.

Ash

The ash content is determined by heating the coal in air at a specified rate, usually up to a temperature of 815°C and this temperature is maintained until constant weight is achieved. The percentage ash content is calculated from the mass of the residue. The above figure is that stipulated in ISO Specifications, although ASTM states a slightly lower figure of 750°C. Ash fusion furnaces are shown in Plates 14 and 15.

The ash emanates from mineral matter present in the coal which would have been derived from soil and clay and any plant remains. A high ash content usually means a lower calorific value which in turn gives rise to problems in combustion, pollution and fly-ash extraction.

Volatile matter

The volatile contents are determined by placing approximately 1 g of a

Plate 14 Ash fusion furnace

Plate 15 Ash muffle furnaces

sample of less than 0.212 mm size in a quartz crucible with a close-fitting lid and placed in an oven for exactly seven minutes at 900°C. The coal is prevented from oxidising because the gases evolved during this process exclude any oxygen. The loss in weight after cooling and reweighing, less the inherent moisture (determined previously) give the volatile content.

It is interesting to note that here the ISO and ASTM methods differ yet again. The above conforms broadly to ISO Specifications, whereas ASTM stipulate the use of platinum crucibles with a temperature of 950°C. It is interesting to note that British and Australian Standards are generally similar to ISO.

Fixed carbon

It is determined by difference, deducting the sum total of the moisture, ash and volatile matter from 100%.

ULTIMATE ANALYSIS

This term covers the determination of carbon, hydrogen, sulphur, nitrogen and oxygen in the coal sample. In practice, it is usually only the sulphur content that is required on a routine basis.

Increasingly, sulphur is determined by means of a Leco analyser in which the sample is subjected to combustion in a high temperature furnace in a plentiful supply of air. The resulting sulphur dioxide produced is quantitatively determined by means of an infra-red detector. This is a very rapid process with results produced by computer printout often within two minutes. This equipment is shown in Plate 16.

There is an alternative high temperature method in which a weighed quantity of coal is burned in a tube furnace at approximately 1,300°C in a stream of oxygen. The sulphur dioxide formed is absorbed in hydrogen peroxide and the result determined volumetrically. A correction has to be made in order to take account of any chlorine which is produced at the same time. This is a longer, more old-fashioned procedure that can be used on occasion to cross-check the Leco result.

CALORIFIC VALUE

In principle, this measures the amount of heat generated when coal is subjected to combustion in a closed steel container under oxygen pressure. The container is called a "bomb" and the heat produced is transferred to water. There are two general ways of determining this value, each involving a different type of calorimeter. The first is by an Adiabatic procedure and the second is by an Isothermal method.

Plate 16 Infra-red sulphur analyser.

In each case the "bomb" is placed in an accurately determined amount of water. When the coal has been subject to combustion, the heat so produced is transferred to the water and the temperature rise of the water accurately measured gives the calorific value which can be calculated to three significant figures. In the case of Adiabatic procedure (namely one in which there is no heat transfer between the water and the bomb), the water container is placed in an environment which adjusts to maintain the same temperature of the water so the amount of heat necessary to maintain such an equilibrium is a measure of the heat transferred to the water and therefore, the calorific value of the coal bringing about such a change can be calculated.

The Isothermal Calorimeter, as the name suggests, is one in which everything maintains the same temperature (usually 25°C) and so the heat generated by the combustion can be directly calculated. Usually, a correction is required for the loss of latent and direct heat in the water vapour and the products of combustion. These values are deducted to give the more practical energy content of the fuel as a net calorific value. However, in the case of the Adiabatic Calorimeter, of course, no corrections of this kind have to be made.

HARDGROVE GRINDABILITY INDEX

This index is of great importance for industrial plants where pulverised coal is fed into boilers. The higher the index, the softer the coal and the easier it is to crush. This results in lower consumption of energy and less wear on crushing elements such as ball mills and wear plates. If the index is too low, it will have the corresponding affect that the grinding plant has too low a capacity for feeding the boiler, with consequently high wear and grinding costs. Another more important factor is that a coal with an inconsistent Hardgrove Index will cause difficulties of control in grinding, combustion, and preheating.

The fact that it is measured by means of an index confirms that this is an empirical entity and can only be evaluated using a standard Hardgrove machine. As it is a comparative procedure, standard reference samples are used for calibration purposes.

In practice, significant variation in results between different laboratories is often due to discrepancies in sample preparation procedures. The sample size required for such a test is normally in the region of 1 kg, which is first air-dried and crushed to pass a 1.18 mm sieve with a minimum amount of material passing the a 0.600 mm sieve.

CRUCIBLE SWELLING NUMBER

This determination has to be carried out in standardised conditions by first heating the sample in a crucible to 820°C. The crucible swelling number is then obtained by comparing the size and shape of the produced button of coke with the criteria shown by standard samples.

This index represents the ability of a given coal to expand during the formation of coke and its main application is for coals designed for the production of coke.

However, it also has an important bearing on steam coal which is purchased for grate furnaces. If the index is too high, an uneven layer of coal will be produced on the grate and consequently will impede the air flow through the layer.

PART IV

Transportation and storage

CHAPTER 17

Problems and dangers

There are two principal issues that can cause serious problems in the transportation and storage of coal. The first, and most dangerous, is spontaneous combustion and the second is frozen coal.

SPONTANEOUS COMBUSTION

As we have established earlier, coal is principally a form of carbon and therefore spontaneous combustion of coal results from the reaction between carbon and oxygen from the atmosphere, in conditions in which the heat generated by the oxidation reaction exceeds the rate at which heat can be dissipated. Therefore, unless some means are established by which the heat produced by the oxidation reaction is dissipated, or that the supply of oxygen is in some way limited, the oxidation of the coal will accelerate and the temperature will increase significantly. The other issue is that oxidation of coal occurs more quickly at higher temperatures and therefore the more heat generated, the faster the oxidation will take place, which in turn will generate even more heat. Obviously, the end product of such a process is combustion.

At normal temperatures, oxidation is a relatively slow process. Consequently, much depends upon the type of coal in question. A coal such as anthracite, which has a low level of volatile matter, will oxidise very slowly. Conversely, coals with high volatile contents oxidise progressively more rapidly.

Obviously, the critical parameter in this process is temperature. One can trace the stages of the reaction according to changes in temperature as follows:

- When the temperature of the coal is in the range of 35–40°C, the coal will oxidise slowly and lose a small amount of its calorific value. For most coals, this loss is minimal (in the order of 1% over a year) assuming that the coal is stored in a compressed state.
- If there is a plentiful supply of oxygen, the coal will absorb the oxygen and the temperature will then rise as oxidation proceeds. Therefore, the temperature will slowly rise to about 50°C.
- When the temperature has risen to approximately 75°C, then this is a warning sign that spontaneous ignition could occur within the next three/four days.
- As oxidation is proceeding more rapidly with the rise in

temperature, it will in turn cause the temperature to rise further, which will then duly increase the rate of oxidation.

- At 140°C carbon dioxide and steam are released (these are the by-products of the oxidation process).
- The oxidation process will then proceed at an increasing rate up to approximately 230°C when the process can be considered to be spontaneous or virtually out of control.
- This leads logically to combustion which occurs usually at around 350°C and, unless carefully controlled, combustion will occur very rapidly, and could get out of control.

This is clearly a most dangerous process which not only results in a significant deterioration in the quality of the coal, but also is a danger to both life and property.

Control procedures

As the chemical processes involved are very simple to identify and as the rate of reaction is so well defined in its link to temperature, it is relatively straightforward to stipulate control procedures which can minimise the chance of spontaneous combustion occurring.

Many of these control procedures are very logical:

- If stockpiles are limited to a maximum height of 8–9 m for small sized coal (from 0–12 mm) and duff coal (coals in the range 0–5 mm), the compression can be minimised. It is also recommended that a maximum height of 6–7 m for graded small grain coal (6–12 mm) should be adequate. For any coals with coarser grain, the maximum stockpile height can be as low as 3–4 m.
- As mentioned above, air supply is very critical in providing oxygen for the oxidation process. Covering a stockpile with a strong plastic cover can be very efficient in limiting the air supply and therefore inhibiting oxidation.
- If the control procedures are ineffective and combustion occurs, one of the most effective counter measures has been the use of dry ice. There are many reported cases of using dry ice to extinguish such fires and in one case 250 kg dry ice per hour was used for extinguishing a fire in a coal store of 25,000 tonnes. In this particular case the dry ice was dug down to approximately half a metre beneath the hearth in order to cover the whole storage area.
- In other examples, fires which occurred deep within a stockpile have been fought by inserting long tubes filled with crushed dry ice deep into the stockpile. Any melted ice is drained and replaced with new dry ice.
- Spontaneous combustion is not a problem limited to coal in stockpiles. If coal (especially fresh and washed coal) is carried in a

ship's hold over long distances, the temperature at which carbon dioxide formation begins to be significant can be little more than 100°C. For this reason the ventilation of holds should only happen above the level of the coal being carried. Any openings which could introduce air into the lower levels of the holds should be sealed.

● It is most important at loading to ensure that the cargo does not become segregated according to particle size. This is usually avoided by trying to ensure that the freefall of the coal is limited, normally by lowering the conveyor belt as far down into the hold as possible.

● Moisture is also an important catalyst for the oxidation process. For example, if dry coal is sprayed, then a certain amount of heat is released. Therefore, if fresh wet coal is loaded in a ship onto older dry coal, heat will be released which will not easily escape and this could increase the temperature to a level at which oxidation will proceed very rapidly with the obvious consequent danger of combustion.

● It is quite clear from the foregoing that all coals with a high level of volatile material must be stored in such a way as to strictly limit the access of air and the consequent ability of fresh air to circulate within the pile. Obviously, it would be even better if such coal could be stored in an inert gas atmosphere.

● One of the most difficult types of material to store is coal that has been newly mined and not subjected to any grading. Consequently, there is a mixture of large lumps and a very high percentage of fines (which could be defined as coal size less than 3 mm). Such coal must be stockpiled very carefully. On the one hand, air should be restricted by stacking the coal in layers and making sure each layer is adequately compressed, otherwise air could have access to the pile and start the oxidation process. However, because of the large proportion of fines, the heat developed will not be able to escape and would bring about a rise in temperature, which would accelerate the oxidation process.

● In some cases, attempts have been made to treat some particular types of coal with oil or tar thus sealing the surface of the coal so that moisture and oxygen cannot interact with the carbon.

● There have been cases where coal has been stored under the sea in order to prevent oxidation.

Storage of coal

Clearly from the foregoing, it is essential that great care is taken when arranging for bulk cargoes of coal to be stored.

When coal is stored in the open, it must be on a firm base which is well drained. The storage area should be so designed as to be secure against flooding. The usual materials used for the base are either asphalt or concrete, or in some cases compressed cinders, coal reject or blast furnace slag.

As with most bulk cargoes, it is imperative that the storage area be always kept clean of materials such as refuse, paper, plastic, wood etc, or even materials such as waste oil which could provide a fire hazard. By the same token, the coal should never be stored near to furnaces, chimneys, steam pipes, hot water tanks or any such constructions that could generate a significant amount of heat. Neither should it ever be stored against brick-work or other porous materials nor over underground drainage channels.

Heat from natural sources such as very strong sunshine can often be underestimated as a hazard for a coal stockpile, but nevertheless this is the case. Therefore, coal should never be stored for excessive periods in very strong sunshine.

The other factor that affects oxidation is air supply and this should also be given consideration when storing coal. If there are any constructions or any factors likely to induce an easy access of air to the stockpile, then these situations should be avoided.

This factor is often very difficult to come to terms with. Coal must either be stored in such a way that the stockpile has good ventilation so that any heat generated can escape, or alternatively, the coal should be so compressed that there is very little air access in order to avoid combustion.

In certain exceptional cases where coal needs to be stored for very long periods, the top and side layers can be compressed and then sealed with a layer of earth. This layer is usually approximately 30 cm thick. In this way the coal can be stored for very long periods.

Stockpile construction

Stockpiles are usually built up by placing and consolidating layers of coal systematically laid by bulldozers or other suitable vehicles. Such layers are of an optimum depth of 15–30 cm. In order to ensure that each layer is evenly spread, harrows are used and the layer then consolidated using rollers.

Normally, the sides of a good stockpile are shaped to produce a uniform slope of not more than 30° to the horizontal.

It is usual to check the compaction of each layer before subsequent layers are put in place. Tools used for this process can also be used to provide a quick temperature check on the compacted material.

Once the stockpile is complete, it should ideally be finished to a convex whaleback form, which has smooth and regular contours. This is usually achieved by a combination of bulldozing and rolling. It is essential that no depressions are left in which rainwater could collect. It is advisable to protect the base of the stockpile by means of a ditch to take any water that could flow into the stockpile area from nearby high ground.

When material has to be taken from a stockpile, it should be done in such a way that the integrity of the stockpiled material is not destroyed. The heap must therefore be left in a condition that the remaining coal is capable of remaining in a safe state for the duration of the stockpiling.

This involves the coal being removed by a scraping process. It is preferable to remove layers of a depth in the range 15–30 cm from the top of the heap. If bulldozers are used, the material scraped away can be pushed well away from the main stockpile.

Under no circumstances should coal be removed using front-end loading shovels as they create dangerous sheer faces and thus destroy the integrity of the stockpile. Such equipment is only suitable if it is the intention to move the entire stockpile, and this must be done within a one-week period.

Stockpile control

When coal is stored, a temperature control must be performed frequently. This is usually performed using thermometers or thermocouples, which are inserted into the stockpile by means of long iron bars placed on the pile with the thermometers and thermocouples in tubes. Such tubes should be driven vertically into the stockpile to a depth of approximately 2 m at regular intervals of 15–20 m on the top of the heap and about 1 m from the edge of the slope. Spacing intervals at the lower end of the range given should be used on the prevailing windward side of the heap.

With stockpiles in excess of 2.5 m in height, it is better to use additional rows of control tubes at each 2.5 m contour line.

The temperature must be read throughout the depth of the tube in order to locate the point of maximum temperature. This is usually between 1 and 1.5 m below the surface.

It is usual to record temperatures on a weekly basis, although this interval will be reduced if temperatures in excess of 35°C are recorded. If, after a period of two months or so there are no signs of heating, the test interval can be extended to two weeks or even monthly.

If steam emission is detected at any stage, it may be necessary to use additional tubes.

Points at which a continuous temperature rise in excess of 35°C is recorded must be investigated and controlled carefully. The critical temperature range is 60–70°C. If temperatures in this range are recorded, the exact extent of coal at that temperature must be located and, if possible, dug out of the pile and spread out in order to allow it to cool.

If any part of a stockpile has begun to burn, the most natural response would be to extinguish the fire using water. However, this would have the effect of flushing away any very fine material and also have the deleterious effect of allowing the access of more air to the stockpile which would in that case exacerbate the fire. The only effective means of extinguishing the fire is to exclude all air, usually by compressing the coal with a bulldozer. When the exterior of the stockpile has reached ambient temperature, the pile can be covered with heavy plastic sheeting in order to ensure that no more air can be admitted.

FROZEN COAL

On occasion, coal cargoes can be sprayed if it is felt there is a danger of excessive dusting due to a high fines content. This means that some coals can be stored with a high surface moisture content. Not surprisingly, such stockpiles can very easily freeze, which makes handling the coal extremely difficult.

Obviously, countries in which severe winter conditions are normal have developed special procedures in order to overcome such difficulties. Most of the procedures are designed to prevent the coal freezing in the first place.

It is self-evident that one of the critical factors concerning the freezing of coal is not only the low ambient temperature, but also the granulometry of the coal as well as the surface moisture content. Relating back to the section on the testing of coal, it is important to note that it is the *surface moisture* and not the hygroscopic moisture that is critical. One of the most effective measures taken to combat freezing has been developed in the USA.

At many of the newer mines, coal is stored in large silos. These silos have ducts designed within the silo walls, through which hot air can be circulated. These silos can hold up to 25,000 tonnes of coal and by means of the hot air ducts the coal can be kept at ambient temperature and thereby freezing is avoided. The hot air supply can often be provided from the waste gases of boiler houses and other similar operations.

Clearly, the problem of freezing coal becomes serious when transportation is involved. In particular, when freezing occurs during the transportation itself. In some cases, pneumatic vibrators have been used in order to separate the coal from ice and (where applicable) snow. It is claimed that this is a more practical and expedient process than endeavouring to defreeze the coal by heating.

Another process of unfreezing coal by heating is the indirect process of heat radiation. This can be effected by using gas burners supplied with reflectors. Also, infra-red radiation has been used to good effect, but like all heating methods suffers from the two major disadvantages of high expense and being extremely slow.

Prevention of freezing

Ideally, coal should be surface-dried as this is the most certain method of avoiding freezing.

If this is not possible, the coal can be treated with chemicals. This has the effect of changing the structure of the ice crystals or effectively reducing the freezing point. Great care must be exercised in deciding which chemicals to use. For example, whereas sodium chloride (common salt) is very effective for treating roads, it should never be used for treating coal because it means that alkaline chloride layers will be created on the surface of boiler tubes, flues and combustion chambers etc.

Calcium chloride in a 10–20% solution will reduce the freezing point to around $-8°C$. The problem is, however, that chemicals usually only have a limited effect because they are at their most effective under ideal conditions. In bad weather conditions such as rain or snow, the chemicals can be washed away and rendered useless.

There is also the danger of chemicals adversely affecting some of the coal testing criteria such as the melting point of the ash. The use of chemicals must, for that reason, be decided upon after very careful tests and consideration.

CHAPTER 18

The role of the cargo superintendent in coal transportation

Many coal producers, because of superior quality control and modern technology, produce excellent coals of consistent quality. One might ask, why the need for a superintendence company to sample every cargo throughout the year? The following fictitious examples will highlight the problems that can occur when loading and discharging coal.

The first example emphasises the importance of keeping a careful control on reclaiming mechanics. Many loading and discharge ports have varied reclaiming mechanisms. One of the most common modes of reclamation is a bucket wheel reclaimer. This is an excellent method and is very manoeuvrable around a stockpile area. The only problem area as far as a superintendence company is concerned, is the height adjustments of the reclaimer in relation to the stockpile floor. Most ports do not have solid concrete bases but varying fill material under the coal stockpiles. Blast furnace slag, granite road base, coal reject material and sand are commonly used. Many loading and discharge ports are very much aware that they must keep a careful control on reclaiming mechanics. The following example emphasises what can happen if control is lost.

The coal producer delivers to the port 40,000 tonnes of premium quality coking coal at 5.8% ash for a specification of 6% ash. For ash results greater than 6%, penalties for every 0.1% ash in excess of 6% are paid. Loading commences on the vessel during heavy rain but the setting on the bucket reclaimer slips and the reclaimer takes an extra 1 m of floor material as well as the coal. The floor material is a mixture of coal wash and slag with an ash of 55%. Loading ends and 2,000 tonnes of coal are still left on the end of the stockpile after 40,000 tonnes of loading.

It is determined that 2,000 tonnes of floor material has been accidentally loaded onto the vessel. The final analysis is carried out and the vessel's ash average is 8%. This might be an extreme case, but it is important to remember that 2,000 tonnes of contamination in many modern ports only takes approximately 30 minutes to load!

The second example illustrates the dangers in thinking in terms of "a stockpile" as meaning "one coal".

Many ports that handle large amounts of different coal types face on-going problems of identification with deliveries by road and rail and stockpile formation. Stockpiles are progressively formed to provide cargo for up-coming vessels. These stockpiles may consist of several different brands of coal being blended for a specification.

Thus, coals delivered to the port may originate from the company loading the vessel, but may not be for that particular stockpile or vessel. The train arrives in the middle of the night shift, has the company name identified at the top of the rail wagon ticket, but the operator fails to see that this train is for another stockpile and vessel. Subsequently, the train is loaded to the wrong stockpile and the resultant blend could be out of specification.

Many coal companies produce several different types of coal ranging from low ash coking coal through to high ash steaming coal. These coals may be delivered to the same port, from the same source, but destined for different stockpiles and different vessels. This example highlights what could happen when there are breakdowns in communications and identification with coal receivals to a port. It is important to remember that many coals look similar but may vary in analysis significantly.

Finally, the ever-present risk of contamination is always on the mind of a diligent cargo inspector.

After the cargo has been successfully delivered to the port, there is another area of potential problem. A nine-hold vessel arrives at the port to load 140,000 tonnes of coal consisting of three different coal types shown in Table 86.

Table 86 Loading sequence for three different coal types

Type	Identification	Tonnes	Ash %	CSN	Holds
1	Coking Coal	80,000	5.8	8	1,3,5,9
2	Steaming Coal	40,000	14.5	1.5	2,4,8
3	Run of Mine Coal	20,000	19.6	1	6,7

The coals are loaded in sequence progressively from the three stockpiles. A cross-contamination of coal, particularly from type 2 or 3 into 1, could affect the final analysis of type 1 markedly. Similarly, when the vessel discharges, the same problem may arise. It should be stressed that modern ports have safeguards to avoid this type of error, but it could still occur.

The three examples outlined are fictitious but feasible and could happen at any time at any port. No systems are guaranteed and can never be. The superintendent's role is to ensure that there is a checking mechanism in case errors of this type occur. This checking mechanism not only protects the coal buyer but also the producer. Coal producers do not wish to jeopardise coal contracts by sending coal to customers that is out of specification. If a problem has occurred, the producer can accurately inform the client what has happened and the long-term relationship between producer and buyer can be maintained.

Coal trade matrix 1980/1985/1990

Coal Trade Matrix 1980 (1,000 tonnes)

Seaborne	Australia	Canada	China	Colombia	Czech	Germany	Poland	S.Africa	UK	USA	USSR	Others*	Total
EC-12: Coking	4,002	548	0	0	139	2,547	4,350	544	35	20,893	1,042	0	34,100
EC-12: Steam	3,182	695	417	0	87	91	9,213	19,676	3,905	12,259	1,605	0	51,130
EC-12: Total	7,184	1,243	417	0	226	2,638	13,563	20,220	3,940	33,152	2,647	0	85,230
Oth Eur: Coking	0	157	0	0	122	0	62	0	0	2,278	402	0	3,021
Oth Eur: Steam	0	35	0	0	11	31	3,538	343	100	237	662	0	4,957
Oth Eur: Total	0	192	0	0	133	31	3,600	343	100	2,515	1,064	0	7,978
N.America: Coking	0	—	0	0	0	0	0	0	0	0	0	0	0
N.America: Steam	60	—	0	0	0	0	235	699	0	0	0	0	994
N.America: Total	60	0	0	0	0	0	235	699	0	0	0	0	994
C&S.America: Coking	60	836	0	200	0	0	975	0	0	4,294	0	0	6,365
C&S.America: Steam	0	0	0	0	0	0	3	0	0	673	0	0	676
C&S.America: Total	60	836	0	200	0	0	978	0	0	4,967	0	0	7,041
Japan: Coking	25,426	10,802	979	—	—	1	387	2,803	—	19,914	1,898	79	62,289
Japan: Steam	3,618	412	1,086	—	—		—	385	—	1,014	233	441	7,189
Japan: Total	29,044	11,214	2,065	0	0	1	387	3,188	0	20,298	2,131	520	69,478
Other Asia: Coking	3,416	1,793	0	0	0	0	0	0	0	875	200	0	6,284
Other Asia: Steam	1,121	0	1,000	0	0	0	9	2,577	0	1,104	0	0	5,811
Other Asia: Total	4,537	1,793	1,000	0	0	0	9	2,577	0	1,979	200	0	12,095
East Europe: Coking	267	0	400	0	0	0	281	0	0	1,761	0	0	2,709
East Europe: Steam	49	0	0	0	0	0	312	0	1	421	0	0	783
East Europe: Total	316	0	400	0	0	0	593	0	1	2,182	0	0	3,492
Africa/ME: Coking	157	0	0	0	0	0	0	0	0	1,388	0	0	1,545
Africa/ME: Steam	0	0	0	0	0	0	0	12	1	56	0	0	69
Africa/ME: Total	157	0	0	0	0	0	0	12	1	1,444	0	0	1,614
Other: Coking	159	0	0	0	0	161	0	0	0	0	0	0	320
Other: Steam	867	0	0	100	0	90	0	88	0	0	0	50	1,195
Other: Total	1,026	0	0	100	0	251	0	88	0	0	0	50	1,515
TOTAL COKING	33,487	14,136	1,379	200	261	2,709	6,055	3,347	35	51,403	3,542	79	116,633
TOTAL STEAM	8,897	1,142	2,503	100	98	212	13,310	23,780	4,007	15,764	2,500	491	72,804
COKING & STEAM	42,384	15,278	3,882	300	359	2,291	19,365	27,127	4,042	67,167	6,042	570	189,437

Source: SS&Y Research Services Ltd, August 1991

*Others includes France, Indonesia, New Zealand, North Korea, Venezuela and Vietnam.

Coal Trade Matrix 1985 (1,000 tonnes)

Seaborne	Australia	Canada	China	Colombia	Czech	Germany	Poland	S.Africa	UK	USA	USSR	Others*	Total
EC-12: Coking	8,884	900	0	1	9	1,600	4,239	256	49	20,520	564	0	37,022
EC-12: Steam	8,418	1,193	236	957	111	389	7,945	25,887	2,228	12,972	880	116	61,332
EC-12: Total	17,302	2,093	236	958	120	1,989	12,184	26,143	2,277	33,492	1,444	116	98,354
Oth Eur: Coking	848	313	0	49	30	0	465	200	0	3,327	2	0	5,234
Oth Eur: Steam	858	56	0	573	0	72	3,932	155	98	1,815	1,261	105	8,925
Oth Eur: Total	1,706	369	0	622	30	72	4,397	355	98	5,142	1,263	105	14,159
N.America: Coking	0	—	0	0	0	0	0	0	0	0	0	0	0
N.America: Steam	63	—	0	539	0	0	0	824	0	0	0	0	1,426
N.America: Total	63	0	0	539	0	0	0	824	0	0	0	0	1,426
C&S.America: Coking	1,114	1,224	0	203	0	0	2,085	0	0	6,331	0	0	10,957
C&S.America: Steam	99	0	0	100	0	14	1	0	0	232	0	0	446
C&S.America: Total	1,213	1,224	0	303	0	14	2,086	0	0	6,563	0	0	11,403
Japan: Coking	29,536	17,026	2,068	—	—	—	—	4,586	—	12,613	2,775	710	69,314
Japan: Steam	14,644	1,516	1,531	—	—	—	—	4,091	—	1,331	1,020	505	24,638
Japan: Total	44,180	18,542	3,599	0	0	0	0	8,677	0	13,944	3,795	1,215	93,952
Other Asia: Coking	7,915	2,879	0	0	0	0	0	300	0	2,302	500	0	13,896
Other Asia: Steam	13,418	1,975	4,035	0	0	0	6	6,975	0	3,597	0	0	30,006
Other Asia: Total	21,333	4,854	4,035	0	0	0	6	7,275	0	5,899	500	0	43,902
East Europe: Coking	726	0	400	0	0	0	1,810	0	0	2,002	0	0	4,938
East Europe: Steam	0	0	0	0	0	0	404	0	1	280	0	276	961
East Europe: Total	726	0	400	0	0	0	2,214	0	1	2,282	0	276	5,899
Africa/ME: Coking	766	31	0	0	0	0	148	0	4	1,260	200	0	2,409
Africa/ME: Steam	577	0	0	100	0	6	6	2,167	174	597	0	0	3,627
Africa/ME: Total	1,343	31	0	100	0	6	154	2,167	178	1,857	200	0	6,036
Other: Coking	0	0	0	0	0	100	0	0	0	0	0	0	100
Other: Steam	0	0	0	231	0	0	0	0	0	0	0	0	235
Other: Total	0	0	0	231	0	100	0	0	4	0	0	0	335
TOTAL COKING	49,789	22,373	2,468	253	39	1,700	8,747	5,342	53	48,355	4,041	710	143,870
TOTAL STEAM	38,077	4,740	5,802	2,500	111	481	12,294	40,099	2,505	20,824	3,161	1,002	131,596
COKING & STEAM	87,866	27,113	8,270	2,753	150	2,181	21,041	45,441	2,558	69,179	7,202	1,712	275,446

Source: SS&Y Research Services Ltd, August 1991
*Others includes France, Indonesia, New Zealand, North Korea, Venezuela and Vietnam.

Coal Trade Matrix 1990 (1,000 tonnes)

Seaborne	Australia	Canada	China	Colombia	Czech	Germany	Poland	S.Africa	UK	USA	USSR	Others*	Total
EC-12: Coking	8,513	2,149	0	0	0	1,251	5,274	195	92	27,804	426	12	45,716
EC-12: Steam	8,590	580	2,720	10,401	1	301	3,111	23,743	1,378	19,492	3,980	686	74,983
EC-12: Total	17,103	2,729	2,720	10,401	1	1,552	8,385	23,938	1,470	47,296	4,406	698	120,699
Oth Eur: Coking	1,946	194	0	0	0	0	625	0	371	2,883	343	65	6,056
Oth Eur: Steam	81	0	62	450	6	50	3,261	1,521	0	113	2,634	452	9,001
Oth Eur: Total	2,027	194	62	450	6	50	3,886	1,521	371	2,996	2,977	517	15,057
N.America: Coking	0	—	0	0	0	0	0	0	0	0	0	0	0
N.America: Steam	22	—	0	1,275	0	0	0	0	0	0	0	323	1,620
N.America: Total	22	0	0	1,275	0	0	0	0	0	0	0	323	1,620
C&S.America: Coking	1,595	1,331	100	0	0	0	2,350	0	0	5,871	0	0	11,247
C&S.America: Steam	177	228	0	73	0	0	0	1,300	0	684	0	200	2,662
C&S.America: Total	1,772	1,559	100	73	0	0	2,350	1,300	0	6,555	0	200	13,900
Japan: Coking	28,187	16,569	1,563	120	—	—	—	3,392	—	10,019	5,682	920	66,452
Japan: Steam	27,054	1,933	2,245	—	—	—	—	1,648	—	2,074	3,022	924	38,900
Japan: Total	55,241	18,502	3,808	120	0	0	0	5,040	0	12,093	8,704	1,844	105,352
Other Asia: Coking	14,099	5,515	1,900	0	0	0	284	0	0	3,286	1,500	81	26,665
Other Asia: Steam	12,693	1,205	8,673	601	0	0	0	14,818	0	4,646	100	579	43,315
Other Asia: Total	26,792	6,720	10,573	601	0	0	284	14,818	0	7,932	1,600	660	69,980
East Europe: Coking	1,643	0	400	0	0	0	230	0	0	2,599	0	0	4,872
East Europe: Steam	33	0	0	0	0	0	43	0	0	47	0	0	123
East Europe: Total	1,676	0	400	0	0	0	273	0	0	2,646	0	0	4,995
Africa/ME: Coking	916	128	0	0	0	0	120	0	0	1,121	150	0	2,435
Africa/ME: Steam	532	0	0	577	0	0	0	2,583	282	1,211	0	0	5,185
Africa/ME: Total	1,448	128	0	577	0	0	120	2,583	282	2,332	150	0	7,620
Other: Coking	0	0	0	0	0	0	0	0	0	1	616	0	617
Other: Steam	0	0	0	102	0	0	0	200	171	0	616	36	1,125
Other: Total	0	0	0	102	0	0	0	200	171	1	1,232	36	1,742
TOTAL COKING	56,899	25,886	3,963	120	0	1,251	8,883	3,587	92	53,584	8,717	1,078	164,060
TOTAL STEAM	49,182	3,946	13,700	13,479	7	351	6,415	45,813	2,202	28,267	10,352	3,200	176,914
COKING & STEAM	106,081	29,832	17,663	13,599	7	1,602	15,298	49,400	2,294	81,851	19,069	4,278	340,974

Source: SS&Y Research Services Ltd, August 1991

*Others include France, Indonesia, New Zealand, North Korea, Venezuela and Vietnam

APPENDIX 2

IMO Code of Safe Practice for Solid Bulk Cargoes*
Appendix B: Coal

Revised during the 31st Session of the Subcommittee on Containers and Cargo, London, January, 1991.

To be issued as a Marine Safety Committee Circular, May 1991.

FOREWORD

Following is a reproduction of the actual draft "Revised schedule for Coal" adopted unanimously by the Subcommittee on Containers and Cargo of the International Maritime Organisation (IMO) at its 31st Session, January 7–11, 1991, in London. A total of 33 nations voted to send the new schedule, which was based on a submission made by the United States government, to the Marine Safety Committee of IMO.

You will note that corrections, deletions and additions are written onto the draft; these are the final changes made by during the Plenary Session of the Subcommittee and they will be incorporated into the final document.

According to IMO procedures, the final document will be presented to the Marine Safety Committee of IMO where, once adopted, it will be issued in the form of a *Marine Safety Committee Circular*, known as an MSC. Once issued as an MSC, the revised schedule will be published in the next edition of the IMO's Code of Safe Practice for Solid Bulk Cargoes.

ANNEX 1. REVISED SCHEDULE FOR COAL

Table: Coal**

BC No.	IMO class	MFAG table No.	Approximate stowage factor m^3/t	EmS No.
010	MHB	311,616***	0.79 to 1.53	B14

*Reproduced with the kind permission of the International Maritime Organisation from the IMO publication *Code of Safe Practice for Solid Bulk Cargoes*.

**For comprehensive information on transport of any material listed, refer to sections 1–9 of this Code.

***Refer to paragraph 6.1.1 (Asphyxia) of the MFAG.

Properties and characteristics

1 Coals may emit methane, a flammable gas. A methane/air mixture containing between 5% and 16% methane constitutes an explosive atmosphere which can be ignited by sparks or naked flame, eg electrical or frictional sparks, a match or lighted cigarette. Methane is lighter than air and may, therefore, accumulate in the upper region of the cargo space or other enclosed spaces. If the cargo space boundaries are not tight, methane can seep through into spaces adjacent to the cargo space.

2 Coals may be subject to oxidation leading to depletion of oxygen and an increase in carbon dioxide in the cargo space.

3 Some coals may be liable to self-heating that could lead to spontaneous combustion in the cargo space. Flammable and toxic gases, including carbon monoxide, may be produced. Carbon monoxide is an odourless gas, slightly lighter than air and has flammable limits in air of 12% to 75% by volume, it is toxic by inhalation with an affinity for blood haemoglobin over 200 times that of oxygen.

4 Some coals may be liable to react with water and produce acids which may cause corrosion. Flammable and toxic gases, including hydrogen, may be produced. Hydrogen is an odourless gas, much lighter than air and has flammable limits in air of 4% to 75% by volume.

Segregation and stowage requirements

1 Boundaries of cargo spaces where materials are carried should be resistant to fire and liquids.

2 Coals should be "separated from" goods of classes 1 (division 1.4), 2, 3, 4, and 5 in packaged form (see The International Maritime Dangerous Goods Code (IMDG Code)) and "separated from" solid bulk materials of classes 4 and 5.1. [Stowage of goods of class 5.1 or solid bulk materials of class 5.1 above or below a coal cargo should be prohibited.][1]

4 Coals should be "separated longitudinally by an intervening complete compartment or hold from" goods of class 1 other than division 1.4
Note: For the interpretation of the segregation terms see paragraph 9.3.3.

General requirements for all coals

1 Prior to loading, the shipper or his appointed agent should provide in writing to the master the characteristics of the cargo and the recommended safe handling procedures for loading and transport of the cargo. As a minimum, the cargo's contract specifications for moisture, sulphur and

---

1. Bracketed sentence above becomes new number 3 under Segregation and stowage requirements.

size should be stated and especially whether the cargo may be liable to emit methane or self-heat.

2. The master should be satisfied that he has received such information prior to accepting the cargo. If the shipper has advised that the cargo is liable to emit methane or self-heat, the master should additionally refer to the Special Precautions.

3. During the loading and while the material remains on board, the master should observe the following:

3.1 All cargo spaces and bilge wells should be clean and dry. Any residue of waste material or previous cargo should be removed, including removable cargo battens, before loading.

3.2 All electrical cables and components situated in cargo spaces and in adjacent spaces should be free of defects. Such cables and electrical components should be safe for use in an explosive atmosphere or positively isolated.

3.3 The ship should carry on board appropriate instruments for measuring the:

3.3.1 concentration of methane in the atmosphere;

3.3.2 concentration of oxygen in the atmosphere;

3.3.3 concentration of carbon monoxide in the atmosphere;

3.3.4 pH value of cargo hold in bilge samples; and

3.3.5 temperature of the cargo in the range between 0°–100°C, without requiring entry into the cargo space.

These instruments should be regularly serviced and calibrated. Ships personnel should be trained in the use of such instruments.

3.4 The ship should carry on board self-contained breathing apparatus required by SOLAS regulation II–2/17. The self-contained breathing apparatus should be worn only by personnel trained in its use.

3.5 Smoking and the use of naked flames should not be permitted in the cargo areas and adjacent spaces and appropriate warning notices should be posted in conspicuous places. Burning, cutting, chipping, welding or other sources of ignition should not be allowed in the vicinity of cargo spaces or in other adjacent spaces, unless the space has been properly ventilated and the methane gas measurements indicate it is safe to do so.

3.6 The master should ensure that the coal cargo is not stowed adjacent to hot areas.

3.7 Prior to departure the master should be satisfied that the surface of the material has been trimmed reasonably level to the boundaries of the cargo space to avoid the formation of gas pockets and to prevent air from permeating the body of the coal. Casings leading into the cargo space should be adequately sealed. The shipper should ensure that the master receives the necessary cooperation from the loading terminal.

3.9   The master should ensure as far as possible that any gases which may be emitted from the materials do not accumulate in adjacent enclosed spaces.

3.8   The atmosphere in the space above the cargo in each cargo space should be regularly monitored for the presence of methane, oxygen and carbon monoxide. Records of these readings should be maintained. The frequency of the testing should depend upon the information provided by the shipper and the information obtained through the analysis of the atmosphere in the cargo space. Means should be provided for testing the atmosphere in the space above the cargo without opening the hatch covers. Where such monitoring indicates the presence of methane, or a rise in temperature or the presence of carbon monoxide, the master should refer to the relevant special precautions.

3.10 The master should ensure that enclosed working spaces, eg store-rooms, carpenter's shop, passage ways, tunnels, etc, are regularly monitored for the presence of methane, oxygen and carbon monoxide. Such spaces should be adequately ventilated.

3.11 A system of regular hold bilge testing should be carried out. If the pH monitoring indicates that a corrosion risk exists, the master should ensure that all hold bilges are kept dry during the voyage in order to avoid possible accumulation of acids on tank tops and in the bilge systems.

3.12 If the behaviour of the cargo during the voyage differs from that specified in the cargo declaration, the master should report such differences to the shipper. Such reports will enable the shipper to maintain records on the behaviour of the coal cargoes, so that the information provided to the master can be reviewed in the light of transport experience.

3.13 The Administration may approve alternative requirements to those recommended in this schedule.

Special precautions

Coals emitting methane

If the shipper has advised that the cargo is liable to emit methane or the analysis of the atmosphere in the cargo space indicates the presence of methane, the following additional precautions should be taken:

1. Adequate surface ventilation should be maintained. On no account should air be directed into the body of the coal as air could promote self-heating.

2. Care should be taken to vent any accumulated gases prior to removal of the hatch covers or other openings for any reason, including unloading. Cargo hatches and other openings should be opened carefully to avoid creating sparks. Smoking and the use of naked flame should be prohibited.

3. Personnel should not be permitted to enter the cargo space or adjacent spaces unless the space has been ventilated and the atmosphere tested and found to be gas-free and has sufficient oxygen to support life. If this is not possible, emergency entry into the space should be undertaken only by trained personnel wearing self-contained breathing apparatus, under the supervision of a responsible officer. In addition, special precautions to ensure that no source of ignition is carried into the space should be observed.

4. The master should ensure that enclosed working spaces, eg. store rooms, carpenter's shops, passage ways, tunnels, etc, are regularly monitored for the presence of methane. Such spaces should be adequately ventilated and, in the case of mechanical ventilation, only equipment safe for use in an explosive atmosphere should be used. Testing is especially important prior to permitting personnel to enter such spaces or energizing equipment within those spaces

Self-heating coals

1. If the shipper has advised that the cargo is liable to self-heat, the master may wish to seek confirmation that the precautions intended to be taken and the procedures intended for monitoring the cargo during the voyage are adequate.

2. If the cargo is liable to self-heat or the analysis of the atmosphere in the cargo space indicates an increasing concentration of carbon monoxide or the temperature of the cargo is rising rapidly, then the following additional precautions should be taken:

2.1 The hatch covers should be closed immediately after completion of loading in each cargo space. The hatch covers can also be additionally sealed with a suitable sealing tape. Surface ventilation should be limited to the extent necessary to remove gases which may have accumulated. Forced ventilation should not be used. [On no account should air be directed into the body of the coal.]

2.2 Personnel should not be allowed to enter the cargo space, unless they are wearing self-contained breathing apparatus and access is critical to the safety of the ship or safety of life. The self-contained breathing apparatus should be worn only by personnel trained in its use.

2.3 When required by the competent authority, the temperature of the cargo in each cargo space should be measured at regular intervals to detect self-heating.

2.4 If the temperature of the cargo exceeds 55°C, and the temperature or the carbon monoxide level is increasing rapidly, a potential fire situation may be developing. The cargo space should be completely closed down and all ventilation ceased. The master should seek expert advice immediately and should consider making for the nearest suitable port of refuge. Water should not be used for cooling the material or fighting coal cargo fires at sea, but may be used for cooling the boundaries of the cargo space.

ANNEX 2. EMERGENCY SCHEDULE B 14 (APPENDIX E)

Coal (BC no. 010)
Special emergency equipment to be carried Nil
EMERGENCY PROCEDURES Nil EMERGENCY ACTION IN A FIRE SITUATION Batten down. Exclusion of air may be sufficient to control the fire. **Do not use water**. Seek expert advice and consider heading for nearest suitable port. **Medical first aid** MFAG table no.: 311, 616 and refer to paragraph 6.1.1. (Asphyxia) of the MFAG

Remarks: The use of CO_2 or inert gas, if available, should be withheld until fire is apparent.

Draft surveying

Further to the description of draft surveying as means of determining the weight of coal cargoes, set out below is the Draft Survey Code of Practice employed by CASCO (Cargo Superintendents Co of Australia).

The accuracy of any survey is only as good as the care taken over all readings and measurements. The operational standards, general condition and hydrostatic particulars vary considerably from ship to ship so that the surveyor must use his experience and judgement in each case and keep records for reference at a later date.

1. Prior to commencing the draft survey, the surveyor should check for any job requirements applicable to that survey.

2. Check and see if there is any previous information available on the vessel. The surveyor should attend the vessel at the earliest opportunity even if commencement of loading or discharging is some time after arrival, as this allows the surveyor time for rechecking whilst avoiding delay to cargo operations.

3. *Drafts* All drafts to be visually read preferably in conjunction with a ship's officer and where possible a boat should be used. Any deviation from this should be recorded. The drafts read should be corrected to the perpendiculars by calculation where applicable.

4. *Calculation* The survey should be calculated on the double mean of means principle using the vessel's displacement scale. First and second trim corrections must be calculated and applied.

5. *Densities* Water density samples to be taken using a "Tumblefill" salinometer pot and the density measured using a glass zeal draft survey hydrometer. For draft surveys density air is used, as opposed to density vacuum as obtained from a loadline hydrometer. The hydrometer should have a certificate of conformity and test or be regularly checked against a duly certified hydrometer. Water samples should be taken at 15, 50 and 85% of the vessel's draft at quarter length forward and aft of midships and the results averaged for the survey. The result obtained should be compared with recent port findings, whilst taking into account past and prevailing weather conditions, the state of the tide and any other factors liable to have an influence on the water density. The surveyor should ensure that he is satisfied with the result obtained. If a strange or unexpected result is found, the densities should be retaken, if necessary at 1 m intervals and checked on both port and starboard sides of the vessel.

6. *Ballast* All soundings of full ballast tanks should be taken with a calibrated steel sounding tape and water finding paste. It is sometimes more accurate to use a rod when tanks are empty as the "bob" on the end of the tape can only be marginally smaller than the pipe diameter. The surveyor must use his judgement.

Density of the water in the ballast tanks should be measured whenever possible. The water density of ballast holds should be checked at 15, 50 and 85% of the depth using a glass zeal draft survey hydrometer and the results averaged. Corrections for trim and heel must be applied from the vessel's surrounding/ullage tank calibration tables. Any tables with poorly presented capacities or corrections should be specifically mentioned on the comments page of the draft survey report. Should there be different books or entries for tanks with more than one set of sounding pipes, this should be mentioned.

It is impossible to cover all aspects of calculating the ballast quantities in all vessels and surveyors must use their experience and judgement to ensure that this area of potential errror is measured and calculated as accurately as possible. Where any doubts exist as to the quantity of remaining ballast in "empty" ballast tanks, then this must be calculated from the ship's tables using the capacity shown against zero if no sounding is obtained.

7. *Fresh water* The quantity of fresh water in the fresh water tanks should be determined either by reading the gauges if fitted or by sounding the tanks and corrections applied if applicable.

8. *Bunkers* The quantity of bunkers on board should be obtained from the chief engineer at the initial survey and allowance made to this figure for port consumption and any bunkers shipped for the final draft survey calculation.

9. *Bilges* All bilges should be sounded at initial and final surveys or confirmed dry. Where a cargo being loaded into a vessel is likely to contain run off moisture, a request should be made by the surveyor to the ship's master asking that a record of bilge soundings and the quantities of bilge water pumped overboard during the voyage be kept and that this record be made available to the receiver's surveyor at the discharge port.

10. *Stores* Any large amounts of stores or spare gear shipped or transhipped should be allowed for in the draft survey calculation.

11. *Survey conditions* Survey conditions must be recorded at both the initial and the final surveys. The following list sets out the conditions to be noted:

- Sea state
- Swell
- Current
- Vessel's motion
- Wind force and direction

- Air temperature
- Sea temperature
- Visibility
- Amount of cargo residue on deck
- Amount of stores loaded, discharged or transhipped
- Tidal times
- Prevailing weather conditions at time of survey
- Weather conditions throughout loading/discharge

12. *Cargo residue* At a survey on completion of discharge, a visual hold inspection should be carried out and a relevant comment on cargo residues incorporated in the report.

13. *Comments* The surveyor should include a paragraph or page of general comments with the draft survey report showing:

(a) Any deviation from this Code of Practice.
(b) Relevant or helpful information pertinent to the ship or the survey.
(c) Areas where care must be taken in interpolation or other use of the ship's tables.
(d) Any other problems encountered in carrying out the draft surveys.

Index